"CHANGES in the economic landscape leave people today with little connection to their industrial heritage because most of the physical elements of this history are beyond anyone's means to save. *Steel: The Diary of a Furnace Worker* is an excellent industrial resource that has been saved, explains this important heritage, and provides an unadulterated account of real life in a steel mill."

—August R. Carlino, President & CEO,
Steel Industry Heritage Corporation

STEEL: the Diary of a Furnace Worker

The Original 1922 Edition by
CHARLES RUMFORD WALKER

With a New Preface, Afterword, Glossary & Photographs

Edited by
KENNETH J. KOBUS

The Iron and Steel Society
Warrendale, PA

The text of *Steel: The Diary of a Furnace Worker* by Charles Rumford Walker was originally published by Atlantic Monthly Press, Boston in 1922.

Annotated edition © 1999 by the Iron & Steel Society.
All rights reserved.
Printed in the U.S.A.

ISBN 1-886362-36-X

Library of Congress Catalog Number (((XXXXXXXXX)))

10 9 8 7 6 5 4 3 2 1

Iron and Steel Society
186 Thorn Hill Road
Warrendale, PA 15086-7528
Tel.: (724) 776-1535
Fax: (724) 776-0430
E-mail: custserv@issource.org
www.issource.org

ABOUT THE COVER

THIS artist's view of "pitside" operations depicts the "tapping" and "teeming" process, probably at the Pittsburgh or the Aliquippa Works of Jones and Laughlin Steel. Both shops had tilting furnaces. This artwork adorned the cover of a company publication in the 1930s. Reprinted with permission of LTV Steel Co., Inc.

IN MEMORY OF

My grandfathers
Vid Salopek and Andrew Kobus

My father
John Kobus

Steelworkers all

Contents

ACKNOWLEDGMENTS		viii
PREFACE		ix
ABOUT CHARLES RUMFORD WALKER		xiii
	FOREWORD	xvii
I	CAMP EUSTIS Bouton, Pennsylvania	1
II	MOLTEN STEEL IN THE "PIT" An Initiation	25
III	THE OPEN-HEARTH FURNACE Night-Shifts	39
IV	EVERYDAY LIFE	57
V	WORKING THE TWENTY-FOUR HOUR SHIFT	71
VI	BLAST-FURNACE APPRENTICESHIP	91
VII	DUST, HEAT, AND COMRADESHIP	105
VIII	I TAKE A DAY OFF	125
IX	"NO CAN LIVE"	137
	EPILOGUE	151
AFTERWORD		163
GLOSSARY		167

Acknowledgments

I would like to thank the following people without whom this project would not have succeeded: David Jackson of Scunthorpe, England, a friend and fellow steel worker, who provided the initial spark in my desire to prove the identity of Bouton; Mark Houser of the *Pittsburgh Tribune-Review*, Louise Sturgess of the Pittsburgh History and Landmarks Foundation, and William Obenchain of the American Iron and Steel Institute, Washington D.C. for their backing and support; fellow J&L/LTV employee Don Inman for his invaluable assistance in securing photographs; and finally, the members and staff of the Iron and Steel Society for sharing a vision.

Ken Kobus
Valparaiso, Indiana
March 1999

Preface

WHEN Charles Walker left the U.S. Army in 1919, he embarked on an idealistic project to explore an "almost equal interest in the process of steelmaking, the administration of business, and the problem of industrial relations" (p. 151), something we could expect of a young, energetic, intelligent, college graduate of the period. By his own admission he "never sought information as an investigator. Most of [his] energy of mind was spent upon doing the job at hand; and what impressions [he] received came unsought in the course of a day's work" (p. 151). But, in his effort to explore the mores of "modern" economic theory, he serendipitously exposes to us a far more basic theme: the feelings of man for man, and in particular for those who, by Walker's own definition, "work at the bottom of society" (p. 4) — the immigrant American steelworker.

The text is full of cases of ethnic prejudice. One example of this is on page (p. 53) where Bill the pit boss, after giving a laborer some instructions which were misunderstood, turns to Walker and says, "Lord! but these Hunkies are dumb." Likewise, when asked if it was worth pursuing a position in the cast house, Dippy tells Walker, "You don't want to work there, only Hunkies work on those jobs, they're too damn dirty and too damn hot for a 'white' man" (p. 118). Or, in another example, we read that "Al, the pit boss came through. He was an *American*" (p. 33). In general, we find that Walker carries the banner of support for the immigrant steelworker, although he sometime lapses into the same style of ethnic discrimination as those around him through his use in the text of the terms "Wop" for Italians, and "Hunkey" when referring to Eastern Europeans. But, we know that Walker begins to understand the difficulties faced by immigrants, particularly with respect to language, when he writes, "Here is this Serbian second-helper bossing his third-helper [Walker] in an unknown tongue, and the latter getting the full emotional experience of the immigrant" (p. 53).

In addition, Walker articulately reveals the difficult conditions under which steelmen of the period labored: hot, strenuous, sometimes dangerous work; ten, twelve, and fourteen hour shifts; six-day weeks; and twice per month, what was known as the "long turn," twenty-four hours of continuous work in order to make the transition from the day shift to the night shift.

Walker's book describes operations at a blast furnace/open hearth/Bessemer converter steelmaking facility in Aliquippa, Pennsylvania. This fact alone is important because neither the Bessemer nor the open hearth process of steelmaking is employed in the United States any longer. Adding to the rarity is the fact that in this plant the open hearth furnaces were of the tilting variety. Relatively few of these shops were built in the U.S. because of the high cost of construction and maintenance. Additionally, most of the Aliquippa facility has been closed and the buildings razed. All that we are left with is this historically significant description of this place, and more importantly its people, our ancestors, some of whom we may not have had the pleasure of meeting.

Though this story is about the Jones and Laughlin Steel Company's Aliquippa Works, nowhere in the book is this mentioned. Walker speaks of working in the Bouton mill and living in the town of Bouton. If you check maps or other references for the time in question, you will find no town named Bouton in the state of Pennsylvania. This leaves us in a quandary: Why would the author hide this important fact? Because he was "keenly interested in economic and social values" (p. 151), Walker's intent was probably to protect his sources, i.e. the superintendents and managers who let him work in the mill. He felt that "management [at the mill] was exceedingly efficient and fair minded" (p. 152) so he would want them to come to no harm. The clues that he left behind, facts that support the real identity of the plant, are explored in the Afterword.

Steel was the basic industry of America. It made the United States a leader in the eyes of the world. Little is written that describes the industry from other than a technological viewpoint. This firsthand account, and others, such as *Out of This Furnace* by Thomas Bell, give us a more complete view of an immigrant worker's life both inside and outside the workplace.

Did Walker's efforts to call attention to needed reforms pay off? Probably not. By the time the book was published in 1922 many of the reforms that he was anticipating had already come to fruition. Many plants had eliminated the "long turn" and the twelve-hour day, while great strides were also being made in safety. Unions, however, would wait for another day.

One point worthy of correction, or at least comment, is the story related by Walker on page 36, of 24 men being killed and buried in a Bessemer accident when molten steel got loose. Nowhere could I find documentation of an incident where 24 people were killed in a year, let alone in one accident at the plant, during the years in question. Mill stories such as this one are usually embellished in each stage of the telling. To put it into context, 25 tons of metal, which would be the amount in a large Bessemer converter, would cover a nine-by-twelve-foot room to a depth of about one foot. An area of this size would be very crowded with twenty-four people in it, and pit areas around furnaces are usually large. I have tried to research and confirm several reported incidents of a similar nature, also unsuccessfully.

A glossary provides explanation of the steelmaking terms Walker uses in the text. Photographs are also included to help give those lacking technical knowledge of the steel industry a better understanding of the steelmaking processes described by Walker. Actual photographs of the Aliquippa plant of Jones and Laughlin and the town of Aliquippa (which was formerly known as Woodlawn) are used whenever possible. Many of the photos are contemporary to Walker's experience.

About Charles Rumford Walker

CHARLES Rumford Walker was born on July 31, 1893 in Concord, New Hampshire, the son of a physician. Educated at Phillips Exeter Academy and Yale University, he received a BA in 1916, and was a member of Phi Beta Kappa and Skull and Bones. In 1917 he entered the army and from September 1918 until February 1919 was a member of the Allied Expeditionary Force. At discharge he had reached the rank of first lieutenant.

For a period of time in 1919, Walker was employed as a manual laborer in the Open Hearth and Blast Furnace Departments of the Aliquippa, Pennsylvania Works of the Jones and Laughlin Steel Company. The experience gained was the basis for his first work, *Steel: The Diary of a Furnace Worker* (1922).

Considered to be an expert in human relations in the metals industries, Walker was a pioneer in the study of technology's impact on human behavior and organizations, and directed the Yale Project on technology and industrial relations. Walker published other works including *Bread and Fire* (1927), a novel about his experience in the brass industry; *Human Relations in an Expanding Company* (co-author, 1948); *Steeltown: An Industrial Case History* (1950); *The Man on the Assembly Line* (co-author, 1952); *The Foreman on the Assembly Line* (co-author, 1956); *Toward the Automatic Factory* (1957); *Modern Technology and Civilization* (1962); and *Technology, Industry and Man: The Age of Acceleration* (1968).

As a result of his work *Steel: The Diary of a Furnace Worker,* Walker was invited to become an assistant editor of *Atlantic Monthly* magazine. Later he served on the staff of the *Independent* magazine, and subsequently as an associate editor of *The Bookman.* In 1938 he held a Guggenheim

Fellowship in history. He returned to Yale in the 1940s, retiring as a Senior Research Fellow in Technology and Society in 1962. From then until his death in 1974, Walker held the title of Curator of Yale Library's Technology and Society Collection, which was given to Yale by Walker upon his retirement.

A man of varied interests, Walker also wrote several plays and was responsible for the production of others. In addition, he translated plays of Euripides, Sophocles, and Aeschylus from the Greek.

He died on November 26, 1974 in Wellfleet, Massachusetts, at the age of 81.

Foreword

IN the summer of 1919, a few weeks before the Great Steel Strike, I bought some second-hand clothes and went to work on an open-hearth furnace near Pittsburgh to learn the steel business. I was a graduate of Yale, and a few weeks before had resigned a commission as first-lieutenant in the regular army. Clean-up man in the pit was my first job, which I held until I passed to third-helper on the open-hearth. Later I worked in the cast-house, became a member of the stove-gang, and at length achieved the semi-skilled job of hot-blast man on the blast furnace. I acquired the current Anglo-Hunky language and knew speedily the grind and the camaraderie of American steel-making. In these chapters I have put down what I saw, felt, and thought as a steel-worker in 1919.

Steel is perhaps the basic industry of America. In a sense it is the industry that props our complex industrial civilization, since it supplies the steel frame, the steel rail, the steel tool without which locomotives and skyscrapers would be impossible. And in America it contains the largest known combination of management and capital, the United States Steel Corporation. Some appreciation of these things I had when I went to work in the steel business. It was clear that steel had become something of a barometer not only for American business but for American labor. I was keenly interested to know what would happen, and believed that basic industries like steel and coal were cast for leading rôles either in the breaking-up or the making-over of society.

FOREWORD

The book is written from a diary of notes put down in the evenings when I was working on day shifts of ten hours. Alternate weeks, I worked the fourteen-hour night shift, and spent my time off eating or asleep.

The book is a narrative — heat, fatigue, rough-house, pay, as they came in an uncharted wave throughout the twenty-four hours.

But it is in a sense raw material, I believe, that suggests the beginnings of several studies both human and economic. Mr. Walter Lippmann has recently pointed out that men do not act in accordance with the facts and forces of the world as it is, but in accordance with the "picture" of it they have in their heads.[1] Nowhere does the form and pressure of the real world differ more sharply from the picture in men's heads than among different social and racial groups in industry. Nor is anywhere the accuracy of the picture of more importance. An open-hearth furnace helper, working the twelve-hour day, and a Boston broker, owning fifty shares of Steel Preferred, hold, as a rule, strikingly different pictures of the same forces and conditions. But what is of greater importance is that director, manager, foreman, by reason of training, interest, or tradition, are often quite as unable to guess at the picture in the worker's head, and hence to understand his actions, as the more distant stock-holder.

Perhaps a technique may some day arise which will supply the executives of industry not only with the facts about employees in their varied racial and social groups, but supply the facts with *due emphasis* and in *three dimensions* so that the controller of power may be able to see them as descriptive of men of like mind with himself. The conclusion most burned

[1] *Public Opinion:* Harcourt, Brace and Company, *1922*

FOREWORD

into my consciousness was the lack of such knowledge or understanding in the steel industry and the imperative need of securing it, in order to escape continual industrial war, and perhaps disaster.

There are certain inferences, I think, like the above, that can be made from this record. But no thesis has been introduced and no argument developed. I have recorded the impressions of a complex environment, putting into words sight, sound, feeling and thought. The book may be read as a story of men and machines and a personal adventure among them no less than as a study of conditions and a system.

C. R. W.

I

CAMP EUSTIS—BOUTON, PENNSYLVANIA

A SMALL torrent of khaki swept on to the ferryboat that was taking troops to the special train for Camp Merritt. They stood all over her deck, in uncomfortably small areas; there seemed to be no room for the pack, which perhaps you were expected to swallow. Faces were a little pale from seasickness, but carried a uniformly radiant expression, which proceeded from a lively anticipation of civilian happiness. The conversation was ejaculatory, and included slapping and digging and squeezing your neighbor. Men were saying over and over again: "This is about the last li'l war they'll ketch me for."

I succeeded in getting beside the civilian pilot.

"What's happening in America?" I asked.

"Oh," he said, "it's a mess over here. There ain't any jobs, and labor is raisin' hell. Everybody that hez a job strikes." He looked out over the water at a tug hurrying past. "I don't know what we're comin' out at. Russia, mebbe."

In the spring of the year Camp Eustis was an island of concrete roads and wooden barracks salvaged from an encroaching sea of mud. Its site had been selected at an immense distance from any village, or even any collection of human dwellings, for particular reasons. It was to contain the longest artillery range in the United States.

After wallowing in bog road through Virginian forest, one came with a shock of relief to a wide, raised, concrete roadbed, which passed newly built warehouses and, after an eighth of a mile, curved into the centre of the camp.

It was like any one of the score of mushroom military centres that grew up on American soil in the years from 1917 to 1919, except that there was an unusual abundance of heavy guns. They covered field upon field, opposite the ordnance warehouses, and their yellow and green camouflage looked absurdly showy in the spring sunshine. Mornings, there was apt to be a captive balloon or two afloat from the balloon school, against blue sky and white clouds; and the landscape held several gaunt observation towers, constructed of steel girders and rising from the forest to a height of seventy-five or eighty feet.

The camp was crowded with returning overseas units, awaiting demobilization and praying earnestly for it day by day, as men pray for pardon.

In a few weeks I should be out of this, going to work somewhere, wearing cits. What a variety of moods the world had split into, from the enormous tension that relaxed on the eleventh of November. Geographically, the training-camp was two thousand miles from the devastations of Europe; and from the new forces that were destroying or renewing civilization, how many more? It seemed like the aftermath of an exciting play that had just been acted; waiting here was like staying to put away properties, and dismiss the actors. It occurred to me that the camp was at least ten thousand miles from America.
There was one consolation in this interminable lingering amid the spring muds and rains of Virginia. Duties were light, and there were a hundred and fifty cavalry horses in the stables, needing exercise. Sometimes we went out on the drill-ground and were taught tricks by an old cavalry officer; or hurdles were set up and we practised jumping our horses. The roads were deeply gutted by spring rains and the pressure of heavy trucks, but there were wood trails good to explore, and interesting objectives like Williamstown or Yorktown. I fell into doing my thinking in the saddle.

Naturally, I wondered about my new job — my civilian job. It was not just an ordinary change from one breadwinning place to another. It was a new job in a world never convertible quite to the one that had kindled the war. It was impossible not to feel that the civilized structure had shaken and disintegrated a bit, or to escape the sense of great powers released. I was unable to decide whether the powers were cast for a rôle of great destruction or of great renewal.

Even in Eustis we received newspapers. The urge and groan of those powers naturally worked into phrases now and then, and even into special tightly worded formulæ. I remember newspaper ejaculations, professorial dissertations, orators' exaggerations: "Capital and labor — Labor in its place — The proletariat — A new order" — and so forth. I felt confused and distrustful in the face of phrases and of the implied doctrines, old and new.

Besides the business of demobilizing the national army, the remaining regular officers and non-coms went into the school of fire, and practised observation of shots over a beautiful relief map of the "Chemin-des-Dames." This was the most warlike thing we did and continued for several months.

One day I took a walk beside the ordnance warehouses, and looked over at the rows of guns stretching for a quarter of a mile beside railroad tracks. In a short time I would be turning my back on these complicated engines. I was even sorry about it, a little; I had spent so much sweat and brain learning about their crankinesses.

In that civil life to follow, I began to see that I wanted two things: 1, a job to give me a living; 2, a chance to discover and build under the new social and economic conditions.

I was twenty-five, a college graduate, a first-lieutenant in the army. In the civilian world into which I was about to jump, most of my connections were with the university I had recently left, few or none in the business world. Why not enlist, then, in one of the basic industries: coal, oil or steel? I liked steel — it was the basic American industry, and technically and economically it interested me. Why not enlist in steel? Get a laborer's job? Learn the business? And besides, the chemical forces of change, I meditated, were at work at the *bottom* of society —

The next day I sent in the resignation of my commission in the regular army of the United States.

Outside the car window, ore piles were visible, black stacks and sooty sheet-iron mills, coal dumps and jagged cuts in the hills against greenness and the meadows and mountains beyond. There were farms, here and there, but they seemed to have been let in by sufferance amid the primary apparatus of the steel-makers.

What an amazingly primary thing steel had become in the civilization we called modern! Steel was the basic industry of America; but more than that, it was, in a sense, the buttress, the essential frame, rather, of present-day life. It made rails, surgical instruments, the girders of skyscrapers, the tools which cut, bored, and filed all the other tools that made, in their turn, the material basis of our living. It was interesting to think that it contained America's biggest "trust," the greatest example of integration, of financial, of managerial combination, anywhere to be found. Steel was critical in America's future, wasn't it — critical for business, critical for labor?

I met a salesman on the train, who was about to go into business for himself. "I intend to start out on a new tack," he said.

He told me briefly his life-story, and how things were forcing him to start a new enterprise, alone. He was very much excited by the idea. He was going to quit his employer, having been with him twenty-nine years.

"I'm getting a new job myself," I said; "I've just got out of the army."

We both fell into silence, and thought of our own separate futures.

What were a young man's chances in American business to-day? I thought of a book I had just been reading called, "The Age of Big Business." In it was the story of the first captains who saw a vision of immense material development, and with the utmost vigor and hardihood pushed on and marked the leading trails. But apparently the affair had been too roughly done, the structure too crudely wrought: machinery jarred, broke, threatened to bring life down in a rusty heap. "No, you are wrong," I fancied the business leader saying; "it is the agitator who, by dwelling on imaginary ills, has stirred up the masses of mankind."

I gazed out of the window at the black mills as we passed them. I was about to learn the steel business. I knew perfectly well that the men who built this basic structure were as hardy and intelligent — no less and no more so, I hazarded — as this new generation of mine. But the job — difficult technical job though it was — appeared too simple in their eyes. "Build up business, and society will take care of itself," they had said. A partial breakdown, a partial revolution had resulted. Perhaps a thoroughgoing revolution threatened. I did n't know.

I knew there was no "solution." There was nothing so neat as that for this multiform condition. But an *adjustment*, a *working arrangement* would be found out, somehow, by my generation. I expected to discover no specific — no formula with ribbons — after working at the bottom of the mill. I did expect to learn something of the practical technique of making steel, and alongside it, — despite, or perhaps because of, an outsider's fresh vision — some sense of the forces getting ready at the bottom of things to make or break society. Both kinds of education were certainly up to my generation.

The train jarred under its brakes, and began to slow down.

"Good luck," I said to the salesman; "I hope you make it all right."

"Good luck," he said.

The train stopped and I found the Bouton station, small and neatly built, of a gray stone, with deeply overhanging roof and Gothicized windows. It seemed unrelated to the rest of the steel community. On the right, across tracks, loomed a dark gathering of stacks arising from irregular acres of sheet-iron roofs. Smoke-columns of various texture, some colored gold from an interior light, streaked the sky immediately above the mill stacks. The town spread itself along a valley and on the sides of encircling hills on my left. In the foreground was Main

"Bouton" Station on the Pittsburgh and Lake Erie Railroad main line, much as it would have looked when Walker arrived. The station is really Aliquippa but the station sign still retains the town's original name of Woodlawn. Pittsburgh is behind you in this view. Provenance unknown. Collection of Don Inman.

Street, with stores and restaurants and a fruit-seller. I went across the street to explore for breakfast.

"Can I look at the job?" I asked.
"Sure," he said, "you can look at the job."
I walked out of the square, brick office of the open-hearth foreman, and lost my way in a maze of railroad tracks, trestles, and small brick shanties, at last pushing inside a blackened sheet-iron shell, the mill. I entered by the side,

The Aliquippa Works charging machine was used to fill (charge) the furnace with scrap and other raw materials. The large rod in the center of the machine (called the peel) was used to grab boxes of scrap, thrust the box through the furnace door, and then rotate, dumping the scrap into the furnace. Aliquippa Works, Jones and Laughlin Steel Company. Photo: Don Inman Collection, Beaver County Industrial Museum, Geneva College, Beaver Falls, PA.

following fierce white lights shining from the half-twilight interior. They seemed immensely brighter than the warm sun in the heavens.

I was first conscious of the blaring mouths of furnaces. There were five of them, and men with shovels in line, marching within a yard, hurling a white gravel down red throats. Two of the men were stripped, and their backs were shiny in the red flare. I tried to feel perfectly at home, but discovered a deep consciousness of being overdressed. My straw hat I could have hurled into a ladle of steel.

Someone yelled, "Watch yourself!" and I looked up, with some horror, to note half the mill moving slowly but resolutely

Opposite view of the charging machine, with a charging box on its peel, standing in front of a fixed open hearth furnace. Aliquippa Works, Jones and Laughlin Steel Company. Photo: Don Inman Collection, Beaver County Industrial Museum, Geneva College, Beaver Falls, PA.

CAMP EUSTIS

onward, bent on my annihilation. I was mistaken. It was the charging-machine, rattling and grinding past furnace No. 7.

The machine is a monster, some forty feet from head to rear, stretching nearly the width of the central open space in the mill. The tracks on which it proceeds go the whole length, in front of all the furnaces. I dodged it, or rather ran from it, toward what appeared open water, but found there more tracks for stumbling. An annoyed whistle lifted itself against the general background of noise. I looked over my shoulder. It relieved me to find a mere locomotive. I knew how to cope with locomotives. It was coming at me leisurely, so I gave it an

View down the charging floor of a modern open hearth shop. The furnaces are on the right. A charging machine and a crane with a ladle of molten iron can be seen in the distance. Homestead (PA) Works, Open Hearth Shop #5, United States Steel Corporation. Photo: Archives of Industrial Society, Wm. J. Gaughan Collection, University of Pittsburgh.

interested inspection before leaving the track. It dragged a cauldron of exaggerated proportions on a car fitted to hold it easily. A dull glow showed from inside, and a swirl of sparks and smoke shot up and lost themselves among girders.

The annoyed whistle recurred. By now the charging affair had lumbered past, was still threatening noisily, but was two furnaces below. I stepped back into the central spaces of the mill.

The foreman had told me to see the melter, Peter Grayson. I asked a short Italian, with a blazing face and weeping eyes, where the melter was.

The short bridge from the open hearth shop, down which Walker traveled; we see the power house that Walker mentions. Aliquippa Works, Jones and Laughlin Steel Company. Photo: Don Inman Collection, Beaver County Industrial Museum, Geneva College, Beaver Falls, PA.

CAMP EUSTIS

He stared hostilely at me.

"Pete Grayson," I said.

"Oh, Pete," he returned; "there!"

I followed his eyes past a pile of coal, along a pipe, up to Pete. He was a Russian, of Atlas build, bent, vast-shouldered, a square head like a box. He was lounging slowly toward me with short steps. Coming into the furnace light, I could see he was an old man with white hair under his cap, and a wooden face which, I was certain, kept a uniform expression in all weathers.

View of Aliquippa Works Bessemer Converter #2 in full "blow." The process was invented by Henry Bessemer in 1856 and bears his name. Air, under pressure, was blown through the bottom of the vessel and through the molten iron to refine it into steel. Aliquippa Works, Jones and Laughlin Steel Company. Photo: Don Inman Collection, Beaver County Industrial Museum, Geneva College, Beaver Falls, PA.

"What does a third-helper do?" I asked, when he came alongside.

Pete spat and turned away, as if the question disgusted him profoundly. But I noticed in a moment that he was giving the matter thought.

We waited two minutes. Finally he said, looking at me, "Why a third-helper has got a hell of a lot to do."

He seemed to regard this quantitative answer as entirely satisfying.

"I know," I said, "but *what* in hell does he do?"

The "pit" area of an open hearth shop. In modern shops the furnaces were raised, so the pit area was the lower level, behind the furnaces. The steel was "tapped" into ladles and poured ("teemed") into ingot molds in the pit. Homestead (PA) Works, Open Hearth Shop #5, United States Steel Corporation. Photo: Archives of Industrial Society, Wm. J. Gaughan Collection, University of Pittsburgh.

CAMP EUSTIS

He again looked at the floor, considered, and spat. "He works around the furnace," he said.

I saw that I should have to accept this as a prospectus. So I began negotiations. "I want a job," I said. "I come from Mr. Towers. Have you got anything now?"

He looked away again and said, "They want a man on the night-shift. Can you come at five?"

My heart leaped a bit at "the night-shift." I thought over the hours-schedule the employment manager had rehearsed: "Five to seven, fourteen hours, on the night-week."

"Yes," I said.

The term "pit" is probably derived from the fact that older open hearth and Bessemer shops tapped their furnaces into holes or pits below the ground level, as we see in this view. Homestead (PA) Works, Open Hearth Shop #3, July 1952, United States Steel Corporation. Photo: Archives of Industrial Society, Wm. J. Gaughan Collection, University of Pittsburgh.

We had just about concluded this verbal contract, when a chorus of "Heows" hit our eardrums. Men make such a sound in a queer, startling, warning way, difficult to describe. I looked around for the charging machine or locomotive, but neither was in range.

"What are they 'Heowing' about?" I thought violently to myself.

But Pete had already grabbed my arm with a hand like a crane-hook. "Want to watch y'self," he said; "get hurt."

I saw what it was, now: the overhead crane, about to carry over our heads a couple of tons of coal in a huge swaying box.

I looked around a little more before I left, trying to organize some meaning into the operations I observed; trying to wonder how it would be to take a shovel and hurl that white gravel into those red throats. I said to myself, "Hell! I guess I can handle it," and thought strongly on the worst things I had known in the army.

The main street (Franklin Avenue) of Aliquippa in the era of Walker's tale. Provenance unknown. Collection of Don Inman.

CAMP EUSTIS

As I stood, a locomotive entered the mill from the other end, and went down the track before the furnaces. It was dragging flat-cars, with iron boxes laid crosswise on them, as big as coffins. I went over and looked carefully at the train load, and at one or two of the boxes. They were filled with irregular shapes of iron, wire coils, bars, weights, sheets, fragments of machines, in short — scrap.

"This is what they eat," I thought, glancing at the glowing doors; "I wonder how many tons a day." I waited till the locomotive came to a shaken stop in front of the middle furnace, then left the mill by the tracks along which it had entered.

An exterior view of the company store in Aliquippa, otherwise known as the Pittsburgh Mercantile Company. Employees, their families and locals usually called it the "Merky." Provenance unknown. Collection of Don Inman.

I followed them out and along a short bridge. A little way to my right was solid ground — the yards, where I had been. Back of Mr. Towers's little office were more mills. I picked out the power house — half a city block. Behind them all were five cone-shaped towers, against the sky, and a little smoke curling over the top — the blast furnaces. Behind me the Bessemer furnace threw off a cloud of fire that had changed while I was in the mill from brown to brownish gold. In front, and to my left, the tracks ran on the edge of a sloping embankment that fell away quickly to a lower level. Fifty yards from the base was the blooming-mill, where the metal was being rolled into great oblong shapes called "blooms." A vague red

An interior view of Aliquippa's "Merky." Provenance unknown. Collection of Don Inman.

CAMP EUSTIS

glow came out of its interior twilights.

Down through the railroad ties on which I walked was open space, twenty feet below. Two workmen were coming out with dinner-buckets. It must be nearly twelve. I had a curiosity to know the arrangement and workings of the dark mill-cellar from which they came.

Turning back on the open-hearth mill, when I had crossed the bridge, I could see that it extended itself, in a sort of gigantic lean-to shelter, over what the melter had called the "pit" There was a crane moving about there, and more centres of light, which I took to be molten steel. I wondered about that area, too, and what sort of work the men did.

Food was also sold in the store, as is seen in this view. Provenance unknown. Collection of Don Inman.

When I reached the end of the track, I thought to myself: "I go to work at five o'clock. How about clothes?"

No one in the mill wore overalls, except carpenters and millwrights, and so on. The helpers on the furnaces were clad in shapeless, baggy, gray affairs for trousers, and shirts were blue or gray, with a rare khaki. Hats were either degraded felts, or those black-visor effects — like locomotive engineers.

The twelve-o'clock whistle blew. A few men had been moving toward the gate slowly for minutes. The whistle sent them at top walking-speed. I stared at them to assure myself as to the correct dress for steel makers.

Main Street began at the tracks and ran straight through the town, mounting the hills as it went. At the railroad end was the Hotel Bouton, where I had breakfasted. Beside it was an Italian fruit store sprawling leisurely over the sidewalk, and a

The "one story movie palace" on Franklin Avenue. Rudolph Valentino and Dorothy Dalton star. Provenance unknown. Collection of Don Inman.

Greek restaurant, one of four. The Greeks monopolized the feeding of Bouton. A block farther, on the right, I ran into a clothing-store, a barber-shop, and two rudimentary department stores. Then, on the same side, a finished city block, looking queer and haughty amid its village companions.

"What 's that?" I asked a strolling, raw-boned Slav.

"Comp'ny store," he said.

I passed a one-story movie "palace," almost concealed behind chromatic advertising, and then the street twisted and I entered the "American quarter." Half a mile of neat, slightly varying brick houses, with lawns fifteen by twenty, and children in such quantity as seriously to menace automobiles.

I looked at the numbers with growing interest, to discover in which I should go to bed tomorrow morning at 7.30. The employment manager had given me the number 343 to try. Here

A view of the "American Quarter," with its "slightly varying brick houses." Provenance unknown. Collection of Don Inman.

it was, on the right, quite like the others, and, I guessed, about twenty minutes from the mill. Calculation of the rising-times for future night-shifts came into my mind.

I was shown the back room on the second floor — a very good room, with a big bed, and two windows.

"You can see our garden," said Mrs. Farrell standing at one of the windows.

I looked out and found the most intensively cultivated twenty-foot plot I had ever seen or imagined. Behind was the back road and a mud cliff. The room seemed a little extravagant for a third-helper, but I took it, in order to have a place for the night, and contracted to pay four dollars a week.

As seen from behind the buildings on Franklin Avenue, the houses of Plan 7, which were built for the Poles, Ukrainians, Slovaks, Serbians, and other Eastern Europeans, climb the opposite side of the valley.
Provenance unknown. Collection of Don Inman.

CAMP EUSTIS

I walked through a street where the prices of clothing were moderate, but where there seemed a dearth of second-hand shops. In one store were green suits, belted, and hung on forms. They had the close-fitting waist, and were marked, "Style Plus Garments: Our Special Price, $15.00." The proprietor, who stood in the doorway, to be handy for collaring the prospective customer, rushed out at me, hands threatening. He was of the prevailing racial type.

"Fix you up wid a dandy suit," he said.

"What I am looking for," I said, "is something second-hand. Do you have any?" I shot this out partly as a check.

"Old man upstairs, fix you up. That door."

I went through that door and up two flights, to a room containing an old man, a sewing machine, and a large table covered with old clothing.

"I'm looking for something for working-clothes," I said; "second-hand coat and pants."

View of Plan 12, which housed Germans, Irish, and English. Provenance unknown. Collection of Don Inman.

Typical boarding house found in Aliquippa during this era. Provenance unknown. Collection of Don Inman.

Slope areas with gardens for workers: "the most intensively cultivated...I had ever seen or imagined." Provenance unknown. Collection of Don Inman.

He lifted a number from the tangled mass of garments and displayed them. They appeared to me too clean, too new, too dressy.

"No," I said, "not that."

He searched again and came up with a highly respectable blue coat, with a mere raveling on one sleeve.

"No," I said, "I'll find one."

I fished very deeply, and caught some green pants, evidently "old" and spattered with white paint on the knees. He hastened to point out the white paint.

I tried to explain that I liked a little white paint on my clothes, but saw I was unconvincing. I finally bought the suit with a sort of violence for two dollars, and left with a sense of fortunate escape.

Now for a hat. Two blocks down the street I found one, somewhat soiled and misshapen.

"I'll take that," I said.

The clerk lifted it, and, when I was fumbling for money, brushed off a vast portion of the dirt, and reshaped it into smooth, luxuriant curves. But still I bought the hat.

"At any rate," I thought, "I can restore the thing."

II

MOLTEN STEEL — AN INITIATION

At four o'clock I put on my paint-spattered pants, the coat with a conspicuous hole near one of the buttons, and my green hat. I climbed the little hill before the gate, among leisurely first arrivals, and found myself attracting no attention whatsoever. I felt for the brass check in my shirt pocket, found it, and rebuttoned the pocket. The guard peered into my face, as if he were going to ask for a pass, but did n't.

I walked the four hundred yards to the open-hearth, and noticed clearly for the first time the yard of the blooming-mill. Here, varied shapes of steel, looking as if they weighed several thousand pounds each, were issuing from the mill on continuous treads, and moving about the yard in a most orderly, but complex manner. Electric cranes were sweeping over the quarter-acre of yard-space, and lifting and piling the steel swiftly and precisely on flat cars.

I entered the open-hearth mill by the tracks that ran close to the furnaces. The mill noises broke on me: a moan and rattle of cranes overhead, fifty-ton ones; the jarring of the train-loads of charge-boxes stopping suddenly in front of Number 4; and minor sounds like chains jangling on being dropped, or gravel swishing out of a box. I was conscious of muscles growing tense in the face of this violent environment, a somewhat artificial and eager calm. I walked with excessive firmness, and felt my personality contracting itself into the mere sense of sight and sound.

I looked for Pete.

"He 's in his shanty — over there," said an American furnace-helper, who was getting into his mill clothes.

I went after Pete's shanty. It was a sheet-iron box, 12 by 12, midway down the floor, near a steel beam. Pete was coming out, buttoning the lower buttons of a blue shirt. He looked through my head and passed me, much as he had passed the steel beam. With two or three steps I moved out and blocked his way. He looked at me, loosened his face, and said very cheerfully: "Hello."

"I 've come to work," I said.

"Here," he said, "you 'll work th' pit t' night. Few days, y' know, get used ter things."

He led the way to some iron stairs, and we went down together into that darkened region under the furnaces, about whose function I had speculated.

The blooming mill yard and electric cranes that Walker passed.
Aliquippa Works, Jones and Laughlin Steel Company. Photo: Don Inman Collection, Beaver County Industrial Museum, Geneva College, Beaver Falls, PA.

MOLTEN STEEL

To the left I could make out tracks. Railroads seem to run through a steel mill from cellar to attic. And at intervals, from above the tracks, torrents of sparks swept into the dark, with now and then a small stream of yellow fire.

We stumbled over bricks, mud, clay, a shovel, and the railroad track. In front of a narrow curtain of molten slag, falling on the floor, we waited for some moments. We were under the middle furnaces, I calculated. Gradually the curtain ceased, and Pete leaped under the hole from which it had come.

"Watch yourself," he said.

I followed him with a broad jump, and a prayer about the falling slag.

We came out into the pit, which had so many bright centers of molten steel that it was lighter than outdoors. I watched Pete's back chiefly, and my own feet. We kept stepping between little chunks of dark slag, which made your feet hot, and close to a bucket, ten feet high, which gave forth smoke. Wheelbarrows we met, with and without men, and metal boxes, as large as wagons, dropped about a dirt floor. We avoided a hole with a fire at its center.

At last, at the edge of the pit, near more tracks, we ran into the pit gang: eight or ten men, leaning on shovels and forks and blinking at the molten metal falling into a huge bucket-like ladle.

"Y' work *here*," said Pete, and moved on.

I remember feeling a half-pleasurable glow as I looked about the strenuous environment, of which I was to become a part — a glow mixed with a touch of anxiety as to what I was up against for the next fourteen hours.

Two of the eight men looked at me and grinned. I grinned back and put on my gloves.

"No. 6 furnace?" I asked, nodding toward the stream.

"Ye-ah," said the man next me.

He was a cleanly built person, in loose corduroy pants, blue shirt open at his neck. Italian.

He grinned with extraordinary friendliness and said, "First night, this place?"

"Yes," I returned.

"Goddam hell of a —— job," he said, very genially.

We both turned to look at the stream again.

For ten minutes we stood and stared. Two men lit cigarettes and sat on a wheelbarrow; four of the others had nodded to me; the other three stared.

I was eager to organize into reasonableness a little of this strenuous process that was going forward with a hiss and a roar about me.

"That's the ladle?" I said, to start things.

"Ye-ah, w'ere yer see metal come, dat's spout, crane tak' him over pour platform, see; pour man mak li'l hole in ladle, fill up moul' — see de moul' on de flat cars?"

The Italian was a professor to me. I got the place named and charted in good shape before the night was out. The pit was an area of perhaps half an acre, with open sides and a roof. Two cranes traversed its entire extent and a railway passed through its outer edge, bearing mammoth moulds, seven feet high above their flat cars. Every furnace protruded a spout, and when the molten steel inside was "cooked," tilted backward slightly and poured into a ladle. A bunch of things happened before that pouring. Men appeared on a narrow platform with a very twisted railing, near the spout, and worked for a time with rods. They prodded up inside, till a tiny stream of fire broke through. Then you could see them start back in the nick of time to escape the deluge of molten steel. The stream in the spout would swell to the circumference of a man's body, and fall into the ladle, that oversized bucket thing, hung conveniently for it by the electric crane. A dizzy tide of sparks accompanied

MOLTEN STEEL

the stream, and shot out quite far into the pit, at times causing men to slap themselves to keep their clothing from breaking out into a blaze. There were always staccato human voices against the mechanical noise, and you distinguished by inflection, whether you heard command, or assent, or warning, or simply the lubrications of profanity.

Steel and slag splashing over a ladle and onto the pit floor. The ladle on the left is for steel; the one on the right is for slag. Homestead (PA) Works, Open Hearth Shop #5, Furnace #65, United States Steel Corporation. Photo: Archives of Industrial Society, Wm. J. Gaughan Collection, University of Pittsburgh.

30 STEEL

As the molten stuff rose toward the top of the ladle, curdling like a gigantic pot of oatmeal, somebody gave a yell, and slowly, by an entirely concealed power, the 250-ton furnace lifted itself erect, and the steel stopped flowing down the spout.

But it splashed and slobbered enormously in the ladle at this juncture; a few hundred pounds ran over the edge to the

A ladle of molten steel being swung away from an open hearth furnace.
Homestead (PA) Works, Open Hearth Shop #5, Dec. 30 1948, United States Steel Corporation. Photo: Archives of Industrial Society, Wm. J. Gaughan Collection, University of Pittsburgh.

MOLTEN STEEL

floor of the pit. This, when it had cooled a little, it would be our job to clean up, separating steel scrap from the slag, and putting it into boxes for remelting.

When a ladle was full, the crane took it gingerly in a sweep of a hundred feet through mid-air, and as Fritz said, the men on the pouring platform released a stopper from a hole in the bottom, to let out the steel. It flowed out in a spurting stream three or four inches thick, into moulds that stood some seven feet high on flat cars.

"Clean off the track on Number 7, an' make it fast," from the pit boss, accompanied by a neat stream of tobacco juice, which began to steam vigorously when it struck the hot slag at his feet.

The "pouring platform." Steel is being "teemed" into ingot molds.
Homestead (PA) Works, Open Hearth Shop #5, United States Steel Corporation. Photo: Archives of Industrial Society, Wm. J. Gaughan Collection, University of Pittsburgh.

We passed through to the other side of the furnaces, by going under Number 6, a bright fall of sparks from the slag-hole just missing the heels of the last man.

"Is n't that dangerous and unnecessary?" I said to myself, angrily. "Why do we have to dodge under that slag-hole?"

We moved in the dark along a track that turned in under Seven, into a region of great heat. Before us was a small hill of partially cooled slag, blocking the track. It was like a tiny volcano, actively fluid in the center, with the edges blackened and hard.

I found out very quickly the why of this mess. The furnace is made to rock forward and spill out a few hundred pounds of the slag that floats on top. A short "buggy" car runs under to catch the flow. But somebody had blundered — no buggy was there when the slag came.

"Get him up queek, and let buggy come back for nex' time," explained an Italian with moustachios, who carried the pick. "Huh, whatze matter goddam first-helper, letta furnace go?" he added angrily. "Lotza work."

This job took us three hours. The Italian went in at once with the pick, and loosened a mass of cinder near one of the rails. Fritz and I followed up with shovels, hurling the stuff away from the tracks.

The slag is light, and you can swing a fat shovelful with ease; but mixed with it are clumps of steel that follow the slag over the furnace doors. It grew hotter as we worked in — three inches of red heat, to a slag cake six inches thick.

"Hose," said someone. The Italian found it in back of the next furnace, and screwed it to a spigot between the two. We became drowned in steam.

We had been at it about an hour and a half, and I was shoveling back loose cinder, with a little speed to get it over with. "Rest yourself," commanded Moustachios. "Lotza time, lotza time."

I leaned on my shovel and found rather mixed feelings rising inside me. I was a little resentful at being told what to do; a little pleased that I was up, at least, to the gang standard; a little in doubt as to whether we ought not to be working harder; but, on the whole, tired enough to dismiss the question and lean on my shovel.

The heat was bad at times (from 120 to 130 degrees when you're right in it, I should guess). It was like constantly sticking your head into the fireplace. When you had a cake or two of newly turned slag, glowing on both sides, you worked like hell to get your pick work done and come out. I found a given amount of work in heat fatigued at three times the rate of the same work in a cooler atmosphere. But it was exciting, at all events, and preferable to monotony.

We used the crowbar and sledge on the harder ledges of the stuff, putting a loose piece under the bar and prying.

When it was well cleared, a puffy switch-engine came out of the dark from the direction of Number 4, and pushed a buggy under the furnace. The engineer was short and jolly-looking, and asked the Italians a few very personal questions in a loud ringing voice. Everyone laughed, and all but Fritz and I undertook a new cheekful of "Honest Scrap." I smoked a Camel and gave Fritz one.

Then Al, the pit boss, came through. He was an American, medium husky, cap on one ear, and spat through his teeth. I guessed that Al somehow wasn't as hard-boiled as he looked, and found later that he was new as a boss. I concluded that he adopted this exterior in imitation of bosses of greater natural gifts in those lines, and to give substance to his authority. He used to be a workman in the tin mill.

"All done? If the son of a — — of a first-helper on the furnace had any brains . . ." and so forth. "Now get through and clean out the goddam mess in front."

We went through, and Fritz used the pick against some very dusty cinder that was entirely cool, and was massed in great piles on the front side of the slag-hole.

"Getta wheelbarrow, *you*."

I started for the wheelbarrow, just the ghost of a resentment rising at being "ordered about" by a "Wop" and then fading out into the difficulties I had in finding the wheelbarrow. Two or three things that day I had been sent for — things whose whereabouts were a closed book. "Where the devil," I muttered to myself, violently disturbed, "are wheelbarrows?" I found one, at last, near the masons under Number 4, and started off.

"Hey, what the hell? what the hell?"

So much for that wheelbarrow.

I found another, behind a box, near Number 8, and pushed it back over mud, slag, scrap, and pipes and things. I never knew before what a bother a wheelbarrow is on an open-hearth pit floor. Only four of us stayed for work under Number 7, a German laborer and I coöperating with shovel and wheelbarrow on the right-hand cinder pile.

We had been digging and hauling an hour, and it was necessary to reach underneath the slag-hole to get at what was left. I always glanced upward for sparks and slag when shoveling, and allowed only my right hand and shovel to pass under. Just as arm and shovel went in for a new lot Fritz yelled, "Watch out!" I pulled back with a frog's leap, and dodged a shaft of fat sparks, spattering on the pit floor. A second later, the sparks became a tiny stream, the size of a finger, and then a torrent of molten slag, the size of an arm. The stuff bounded and splashed vigorously when it struck the ground.

It did n't get us, and in a second we both laughed from a safe distance.

"Goddam slag come queek," said Fritz, grinning. "How you like job?" he added.

Before I had any chance to discuss the nuances of a clean-up's walk in life, Fritz was pointing out a new source of molten danger.

We were standing now in the main pit, beyond the overhanging edge of the furnace.

"Look out now, zee!" said Fritz, pointing upward.
Almost over our head was Number 7's spout, and, dribbling off the end, another small rope of sparks.

We fell over each other to the pit's edge, stopping when we reached tracks. Looking back at once, we saw that the stream had thickened like the other in the slag-hole. But here it was molten steel, and with a long drop of thirty feet. The rebound of the thudding molten metal sent it off twenty-five or thirty feet in all directions. Three different groups of men were backing off toward the edge of the pit.

The stream swelled steadily till it reached the circumference of a man's body, and fell in a thudding shaft of metallic flame to the pit's floor. Spatterings went out in a moderately symmetrical circle forty feet across. The smaller gobs of molten stuff made minor centers of spatter of their own. It was a spectacle that burned easily into memory.

The gang of men at the edge of the pit watched the thing with apparent enjoyment. I wondered slowly two things: one, whether anyone ever got caught under such a molten Niagara, and two, whether the pit was going to have a steel floor before it could be stopped. How could it be stopped, anyway?

The craneman had been busy for some minutes picking up a ladle from Number 4, and at that instant he swung it under, and the process of steel-flooring ceased.[1]

What the devil had happened? I talked with everybody I could as they broke up at the pit's edge. It was a rare thing I

---

[1] I learned later the flow could have been stopped by simply tilting back the furnace, but the craneman was ready and so brought the ladle up.

learned: the mud and dolomite (a limestone substance) in the tap-hole had not been properly packed, and broke through. My companions told me about another occasion, some years before, when molten steel got loose. It happened on the Bessemer furnaces, and the workers had n't either the luck or agility of ourselves. It caught twenty-four men in the flow — killed and buried them. The company, with a sense of the proprieties, waited until the families of the men moved before putting the scrap, which contained them, back into the furnace for remelting.

As I ate three bowls of oatmeal at the Greek's, at 7.15, I thought, "Those fellows do these shifts, year after year. What does the heat, and the danger, and the work do to them? Maybe they 'get used to' the whole business. Will I?"

I went to bed at 8.05, and all impressions faded from consciousness, except weariness, and lame arms, and a burn on each ankle.

After two or three days in the pit, I began to know the gang a little by name and character. There was Marco, a young Croat of twenty-four, who had started to teach me Croatian in return for some necessary American; Fritz, a German with the Wanderlust; Adam, an aristocratic person, very mature, and with branching moustachios; Peter, a Russian of infinite good-nature; and a quiet-eyed Pole, who was saving up two hundred dollars to go to the old country.

For several days it was impossible to break into Adam's circle of friends; he would talk and work only with veteran clean-ups, and showed immense pomposity in a knowing way of hooking up slag and scrap to the crane. One day, however, I found him working alone with a wheelbarrow, cleaning cinder from around a buggy car under furnace No. 8. He looked over at me as I passed, and yelled: "Hey, you!"

He wanted my assistance on the wheelbarrow. We worked together for an hour or so, and I felt that perhaps the ice was broken.

"Did you ever work on the floor?" I asked.

"Two years," he said; "no good."

A little later I talked to Marco about him.

"Hell," he said, "he got fired from furnace, for too goddam lazy." I felt less hurt at his snobbishness after that.

Marco and I became good chums. We sat on a wheelbarrow one day, after finishing a job on the track under Six.
"You teach me American," he said; "I teach you Croatian."

"Damn right," I said; and we began on the parts of our body, and the clothing we wore, drawing out some of the words in the dirt with a stick, or marking them with charcoal on a board.

"Did you ever go to school in America?" I asked.

"Three month, night school, Pittsburgh. Too much, work all day, twelve hour, go to school night," he said.

"Do you save any money? Got any in the bank?" I asked, feeling a little fatherly, and wondering on the state of his economic virtues.

"Hell, no," he said; "I don' want money in bank, jes nuff get along on."

I talked to a good many on the savings question, and found the young men very often did n't save, but "bummed round," while practically all the "Hunkies" of twenty-eight or thirty and over saved very successfully. A German who put scrap in the charge-boxes, after the magnet had dropped it, had saved $4000 and invested it. One man said to me: "A good job, save money, work all time, go home, sleep, no spend." Speaking of the German, "He no drink, no spend." The savers, I think, are apt to be the single men who return to their own country in ten or fifteen years.

I came out of the mill one morning after a night-shift, with an appetite that made me run from the railroad bridge to Main Street. I went to the Hotel Bouton where the second-helper on Eight usually eats, and started at the beginning, with pears. I ate the cereal, eggs, potatoes, toast, coffee, and griddle-cakes, taking seconds and thirds when I could negotiate them — the Bouton is stingy under a new management, probably finding that steel-workers eat up the profit. I got up from the table feeling as hungry as when I sat down, and went to the restaurant just two doors below — unpalatable, but serving fairly large portions. There I had another breakfast: coffee, oatmeal, eggs. I felt decidedly better after that, and started home in good humor. But by the time I reached the window of Tom, the Wiener man, I felt that there was room for improvement, and looked in my pocketbook to see if I had any breakfast money left. I had n't a cent, but there were quantities of two-cent stamps. I went in and sat down at Tom's counter, where I ate a bowl of cereal and a glass of milk. Then I opened my purse. In a moment or two I convinced Tom that two-cent stamps were good legal tender, and went home.

III

THE OPEN-HEARTH — NIGHT-SHIFTS

"Have a cigarette, Pete," I said, offering a Camel to a very fat and boyish-looking Russian.

"No t'ank."

"What, no smoke?" I asked, incredulous.

"No, no smoke."

"No drink?" I asked, wondering if I had found a Puritan.

"Oh, *drink*," he said with profound emphasis; and continuing, he told me of other solaces he found in this mortal life.

"Look!" cried some one.

Herb, the craneman, in a fit of extreme playfulness had thrown some wet green paint forty feet through the air at the pit boss, greening the whole side of his face. Al was doing a long backward dodge, and slapping a hand to his painted face, supposing it a draught of hot metal. When he perceived that he was n't killed, he picked up cinder-hunks and bombarded the crane-box. It sounded like hail on tin.

Pete, the Russian melter, came out on the gallery behind the furnaces, and I could see by the way he looked the pit over, that he was picking a man for furnace work. Somebody had stayed out and they were short a helper. He looked at the fat workman beside me, and then grunted.

This was the third time he had picked Russians in preference to the rest of us, who are Serbian, Austrian and American.

The next day I went on the floor, and tackled Pete.

"How about a chance on the floor?" I said, standing in front of him to keep him from lurching away.

"Y' get chance 'nuff, don't worry."

"If I can't get a crack at learning this game in Bouton, I'll go somewhere I can," I said, boiling up a little.

Dick Reber, the Pennsylvania-Dutch melter, came up.

"I want a chance on the floor," I said.

"All right, boy, go on Number 7 to-day."

I made all speed to Number 7. "Is he doing that," I thought, as I picked up my shovel, "because I 'm an American?"

I looked up and saw the big ladle-bucket pouring hot metal into a spout in the furnace-door, accompanied by a great swirl of sparks and flame, spurting upward with a sizzle.

"At last," I said, "I 'm going to make steel."

The steel starts in as "scrap" in the mill-yard. Scrap from anywhere in America; a broken casting, the size of a man's trunk, down to corroded pipe, or strips the thickness of your nail, salvaged in bales. The overhead crane gathers them all from arriving flat cars by a magnet as big as a cart wheel, and the pieces of steel leap to meet it with apparent joy, stick stoutly for a moment, and fall released into iron charge-boxes. By trainloads they pass out of the stock-yard and into the mill, where the track runs directly in front of the furnace-doors. There the charging-machine dumps them quickly into the belly of the furnace. It does its work with a single iron finger, about ten feet long and nearly a foot thick, lifting the box by a cleat on the end, and poking it swiftly into the flaming door. Old furnaces charged by hand, hold from twenty-five to thirty-five tons; new ones, up to two hundred and fifty.

That is the first step in starting to make a "heat," which means cook a bellyful to the proper temperature for steel, ready to tap into a ladle for ingot-making. Next comes "making front-wall," which signifies that no self-respecting brick, clay, or any other substance, can stand a load of metal up to steel-heat without being temporarily relined right away for the next draft of flame. We do that relining by shoveling dolomite into the furnace. The official known as second-helper wields a

THE OPEN-HEARTH

Brobdingnag spoon, about two inches larger than a dinner-plate and fifteen feet long, which a couple of third-helpers, among them myself, fill with dolomite. By use of the spoon, he carefully spreads the protection over the front-wall.

But the sporting job on the open-hearth comes a bit later, and consists in "making back-wall." Then all the men on the furnace and all the men on your neighbor's furnace form a dolomite line, and marching in file to the open door, fling their shovelfuls across the flaming void to the back-wall. It's not a beginner's job. You must swing your weapon through a wide

"Making a back wall." "There all the men on your furnace and all the men on your neighbor's furnace form a dolomite line, and marching in file to the open door, fling their shovelfuls across the flaming void." Homestead (PA) Works, Open Hearth Shop #4, Mar. 1952, United States Steel Corporation. Photo: Archives of Industrial Society, Wm. J. Gaughan Collection, University of Pittsburgh.

arc, to give it "wing," and the stuff must hop off just behind the furnace-door and rise high enough to top the scrap between, and land high. I say it's not a beginner's job, though it 's like golf — the first shovelful may be a winner. What lends life to the sport is the fact that everybody 's in it — it's the team play of the open-hearth, like a house-raising in the community.

A first helper "looking through peepholes in her furnace-doors."
Homestead (PA) Works, Open Hearth Shop #5, United States Steel Corporation. Photo: Archives of Industrial Society, Wm. J. Gaughan Collection, University of Pittsburgh.

THE OPEN-HEARTH

Another thing giving life is the heat. The mouth of the furnace gapes it widest, and you must hug close in order to get the stuff across. Every man has deeply smoked glasses on his nose when he faces the furnace. He's got to stare down her throat, to watch where the dolomite lands. It's up to him to place his stuff — the line is n't marching through the heat to warm its hands. Here's a tip I did n't "savvy" on my first back-wall. Throw your left arm high at the end of your arc, and in front of your face; it will cut the heat an instant, and allow you to see if

The first helper "inspects her brew." This is a rare photo of a "heat" of molten steel "cooking" inside of one of Aliquippa's open hearth furnaces. Aliquippa Works, Open Hearth Furnace #47, Jones and Laughlin Steel Company. Photo: Don Inman Collection, Beaver County Industrial Museum, Geneva College, Beaver Falls, PA.

you have "placed" without flinching. It's really not brawn, — making back-wall, — but a nimble swing and a good eye, and the art of not minding heat.

After that is done, she can cook for a while and needs only watching. The first-helper gives her that, passing up and down every few minutes to look through the peepholes in her furnace-doors. He puts his glasses down on his nose, inspects the brew, and notices if her stomach's in good shape. If the bricks get as red as the gas flame, she's burning the living lining out of her. But he keeps the gas blowing in her ends, as hot as she'll stand

Getting a "bucket of iron" for the open hearth. Here a "torpedo" ladle from the blast furnaces is pouring iron into a ladle for charging into the open hearth furnace, bypassing the mixer. Homestead (PA) Works, Open Hearth Shop #5, United States Steel Corporation. Photo: Archives of Industrial Society, Wm. J. Gaughan Collection, University of Pittsburgh.

THE OPEN-HEARTH

it without a holler. On either end the gas, and on top of it the air. The first-helper, who is cook of the furnace, makes a proper mixture out of them. The hotter he can let the gas through, the quicker the brew is cooked, and the more "tonnage" he'll make that week.

"Get me thirty thousand pounds," said the first-helper when I was on the furnace that first night. Fifteen tons of molten metal! I was undecided whether to bring it in a dipper or in my hat. But it's no more than running upstairs for a handkerchief in the bureau. You climb to a platform near the blower, where the stuff is made, and find a man there with a book. Punch him in the arm and say, "Thirty thou' for Number 7." He will swear moderately and blow a whistle. You return to the furnace, and

Charging iron into a tilting open hearth furnace at the Aliquippa plant.
Aliquippa Works, Open Hearth #44, Jones and Laughlin Steel Company. Photo: Don Inman Collection, Beaver County Industrial Museum, Geneva College, Beaver Falls, PA.

on your heels follows a locomotive dragging a bucket, the ladle, ten feet high. Out of it arise the fumes of your fifteen tons of hot metal. The overhead crane picks it up and pours it through a spout into the furnace. As it goes in, you stand and direct the pouring. The craneman, as he tilts or raises the bucket, watches you for directions, and you stand and make gentle motions with one hand, thus easily and simply controlling the flux of the fifteen tons. That part of the job always pleased me. It was like modeling Niagara with a wave of the hand. Sometimes he

Pouring a test, in this case a "killed" test. In molten steel, expanding gasses make the volume of steel increase or "grow." When aluminum or similar material was added it combined with the gasses and stopped the "growing" process; at this point, the steel is said to be "killed." Homestead (PA) Works, Open Hearth Shop #5, Furnace #65, Dec. 30 1948, United States Steel Corporation. Photo: Archives of Industrial Society, Wm. J. Gaughan Collection, University of Pittsburgh.

THE OPEN-HEARTH 47

spills a little, and there is a vortex of sparks, and much molten metal in front of the door to step on.

She cooks in anywhere from ten hours to twenty-four. The record on this floor is ten, which was put over by Jock. He has worked on most of the open-hearths, I learn, from Scotland to Colorado.

When it's time for a test, the first-helper will take a spoon about the size of your hand and scoop up some of the soup. But not to taste. He pours it into a mould, and when the little ingot is cool, breaks it with a sledge. Everyone on the furnace, barring

The tapping of a tilting open hearth furnace at the Aliquippa Works. Note that the furnace is tipped backward. To remove only slag you would tip the furnace forward. Aliquippa Works, Open Hearth #58, Jones and Laughlin Steel Company. Photo: Don Inman Collection, Beaver County Industrial Museum, Geneva College, Beaver Falls, PA.

myself, looks at the broken metal and gives a wise smile. I'm not enough of a cook. They know by the grain if she has too much carbon or needs more, or is ready to tap, or is n't. With too much carbon, she'll need a "jigger," which is a few more tons of hot metal, to thin her out.

That's about the whole game — abbreviated — up to tap-time. It takes, on an average, eighteen hours, and your shift may be anything from ten to twenty-four. Of course, there are details, like shoveling in fluor-spar to thin out the slag. Be sure you clear the breast of the furnace with your shovelful, when you put that into her. Spar eats the dolomite as mice eat cheese. At intervals the first-helper tilts the whole furnace forward, and she runs out at the doors, which is to drain off the slag that floats on top of the brew. But after much weariness it's tap-time and the "big boss" comes to supervise.

Move aside the shutters covering the round peep-holes on her doors, at this time, and you'll see the brew bubbling away like malt breakfast-food ready to eat. But there's a lot of testing before serving. When it is ready, you run to the place where you hid your little flat manganese shovel and take it to the gallery back of the furnace, near the tap-spout. There you can look down on the pit strewn with those giant bucket-ladles and sprinkled with the clean-up men, who gather painfully all that's spilled or slobbered of hot metal, and save it for a second melting. The whole is swept by the omnipresent crane.

At a proper and chosen instant, the senior melter shouts, "Heow!" and the great furnace rolls on its side on a pair of mammoth rockers, and points a clay spout into the ladle held for it by the crane. Before the hot soup comes rushing, the second-helper has to "ravel her out." That function of his almost destroyed my ambition to learn the steel business. Raveling is poking a pointed rod up the tap-spout, till the stopping is prodded away. You never know when the desired but terrific result is

THE OPEN-HEARTH

accomplished. When it is, he retires as you would from an exploding oil-well. The brew is loose. It comes out, red and hurling flame. Into the ladle it falls with a hiss and a terrifying "splunch." The first and second-helpers immediately make matters worse. They stagger up with bags (containing fine anthracite) and drop them into the mess. They have a most damning effect. The flames hit the roof of the pit, and sway and curl angrily along the frail platform on which you stand. Some occult reasoning tells them how many of these bags to drop in, whether to make a conflagration or a moderate house-burning.

The melter waits a few minutes and then shouts your cue. You and another helper run swiftly along the gallery to the side of the

Two men shoveling manganese into a ladle as Walker describes. This adjusted the chemistry of the steel to desired proportions. In the U.S. these workers would be the second and third helpers. Appleby-Froddingham Steel Company, Scunthorpe, England. Photo: Collection of David Jackson.

spout. At your feet is a pile of manganese, one of the heaviest substances in the world, and seeming heavier than that. It's your job and your helper's to put the pile into the cauldron. And you do it with all manner of speed. The tap stream — at steel heat — is three feet from your face, and gas and sparks come up as the stream hits the ladle. You're expected to get it in fast. You do.

There are almost always two ladles to fill, but you have a "spell" between. When she's tapped, you pick up a piece of sheet iron and cover the spout with it. That's another job to warm frost-bitten fingers. Use gloves and wet burlap — it preserves the hands for future use.

One more step, and the brew is an ingot. There are several tracks entering the pit, and at proper seasons a train of cars swings in, bringing the upright ingot moulds. They stand about seven feet high from their flats. When the ladle is full and slobbering a bit, the craneman swings her gingerly over the first mould. Level with the ladle's base, and above the train of moulds, runs the pouring platform, on which the ingot-men stand.

By means of rods a stopper is released from a small hole in the bottom of the ladle. In a few seconds the stream fills a mould, and the attendant shuts off the steel like a boy at a spigot. The ladle swings gently down the line, and the proper measure of metallic flame squirts into each mould. A trainload of steel is poured in a few minutes.

But this is when all omens are propitious. It's when the stopper-man has made no mistakes. But when rods jam and the stopper won't stop, watch your step, and cover your face. That fierce little stream keeps coming, and nothing that the desperate men on the pouring platform can do seems likely to stem it. Soon one mould is full. But the ladle continues to pour, with twenty tons of steel to go. It can't be allowed to make a steel floor for the pit. It must get into those moulds.

THE OPEN-HEARTH

So the craneman swings her on to the next mould, with the stream aspurt. It's like taking water from the teakettle to the sink with a punctured dipper; half goes on the kitchen floor. But the spattering of molten metal is much more exciting. A few little clots affect the flesh like hot bullets. So, when the crane-man gets ready to swing the little stream down the line, the workers on the platform behave like frightened fishes in a mill pond. Then, while the mould fills, they come back to throw certain ingredients into the cooling metal.

These ingots, when they come from the moulds virgin steel, are impressive things — especially on the night turn. Then each

"Stripped" ingots; the molds have been removed. The ingots are on the way to the blooming mill. Aliquippa Works, Jones and Laughlin Steel Company. Photo: Don Inman Collection, Beaver County Industrial Museum, Geneva College, Beaver Falls, PA.

stands up against the night air like a massive monument of hardened fire. Pass near them, and see what colossal radiators of heat they are. Trainloads of them pass daily out of the pit to the blooming-mill, to catch their first transformation. But my spell with them is done.

I stood behind the furnace near the spout, which still spread a wave of heat about it, and Nick, the second-helper, beside me yelling things in Anglo-Serbian, into my face. He was a loose-limbed, sallow-faced Serbian, with black hair under a green-visored cap, always on the back of his head. His shirt was torn on both sleeves and open nearly to his waist, and in the uncertain lights of the mill his chest and abdomen shone with sweat.

"Goddam you, what you think. Get me" — a long blur of Serbian, here — "spout, quick mak a" — more Serbian with tremendous volume of voice — "furnace, see? You get that goddam mud!"

When a man says that to you with profound emotion, it seems insulting, to say, "What" to it. But that was what I did.

"All right, all right," he said; "what the hell, me get myself, all the work" — blurred here — "son of a — third-helper — wheelbarrow, why don' you — — *quick now when I say!*"

"All right, all right, I'll do it," I said, and went away. I was never in my life so much impressed with the necessity of *doing it*. His language and gesture had been profoundly expressive — of what? I tried to concentrate on the phrases that seeped through emotion and Serbian into English. "Wheelbarrow" — hang on to that; "mud" — that's easy: a wheelbarrow of mud. Good!

I got it at the other end of the mill — opposite Number 4. "Hey! don't use that shovel for mud!" said the second-helper on Number 4.

So I did n't.

I wheeled back to the gallery behind Seven, and found Nick coming out at me. When he saw that hard-won mud of mine, I thought he was going to snap the cords in his throat.

"Goddam it!" he said, when articulation returned, "I tell you, get wheelbarrow dolomite, and half-wheelbarrow clay, and pail of water, and look what you bring, goddam it!"

So that was it — he probably said pail of water with his feet.

"Oh, all right," I said, smiling like a skull; "I thought you said mud. I 'll get it, I 'll get it."

This is amusing enough on the first day; you can go off and laugh in a superior way to yourself about the queer words the foreigners use. But after seven days of it, fourteen hours each, it gets under the skin, it burns along the nerves, as the furnace heat burns along the arms when you make back-wall. It suddenly occurred to me one day, after someone had bawled me out picturesquely for not knowing where something was that I had never heard of, that this was what every immigrant Hunky endured; it was a matter of language largely, of understanding, of knowing the names of things, the uses of things, the language of the boss. Here was this Serbian second-helper bossing his third-helper largely in an unknown tongue, and the latter getting the full emotional experience of the immigrant. I thought of Bill, the pit boss, telling a Hunky to do a clean-up job for him; and when the Hunky said, "What?" he turned to me and said: "Lord! but these Hunkies are dumb."

Most of the false starts, waste motion, misunderstandings, fights, burnings, accidents, nerve-wrack, and desperation of soul would fall away if there were understanding — a common language, of mind as well as tongue.

But then, I thought, all this may be because I'm oversensitive. I had this qualm till one day I met Jack. He was an old regular-army sergeant, a man about thirty. He had come back from fixing a bad spout. They had sledged it out — sledged through the steel that had crept into the dolomite and closed the tap-hole.

"Do you ever feel low?" he said, sitting down on the back of a shovel. "Every once'n while I feel like telling 'em to take their job and go to hell with it; you strain your guts out, and then they swear at you."

"I sometimes feel like a worm," I said, "with no right to be living any way, or so mad I want to lick the bosses and the president."

"If you were first-helper, it wouldn't be so bad," he mused; "you wouldn't have to bring up that damn manganese in a wheelbarrow — and they wouldn't kick you round so much."

"Will I ever get that job?"

We were washing up at one end of the mill, near the Bessemers. There was plenty of hot water, and good broad sinks. I took off my shirt and threw it on top of a locker; the cinder on the front and sleeves had become mud.

Forty men stood up to the sinks, also with their shirts off, their arms and faces and bodies covered with soap, and saying: "Ah, ooh," and "ffu," with the other noises a man makes when getting clean. Every now and then somebody would look into a three-cornered fragment of looking-glass on one of the lockers, and return to apply soap and a scrubbing brush to the bridge of his nose.

A group of Slovene boys, who worked on the Bessemer, picked on one of their number, and covered him with soap and American oaths. Somebody told an obscene story loudly in broken English.

The men who had had a long turn or a hard one washed up silently, except for excessive outbreaks if anybody took their soap. Some few hurried, and left grease or soot on their hands or under their eyes.

"I wash up a little here," said Fred, the American first-helper on Number 7, "and the rest at home. Once after a twenty-four hour shift, I fell asleep in the bathtub, and woke up to find the water cold. Of course, you can't really get this stuff off in one or two wash-ups. It gets under your skin. When the furnace used to get down for repairs, and we were laid off, I'd be clean at the end of a week." He laughed and went off.

I had scraped most of the soot from arms and chest, and was struggling desperately, with the small of my back. A thick-chested workman at the next bowl, with fringes of gray hair, and a scar on his cheek, grabbed the brush out of my hand.

"Me show you how we do in coal-mine," he said; and proceeded vigorously to grind the bristles into my back, and get up a tremendous lather, that dripped down on my trousers to the floor.

"You wash your buddy's back, buddy wash yours," he said.

I went out of the open-hearth shelter slowly, and watched the line — nearly a quarter of a mile long — of swinging dinner-buckets. Some were large and round, and had a place on top for coffee; some were circular and long; some were flat and square. I looked at the men. They were the day-shift coming in.

"I have finished," I said to myself automatically. "I'm going to eat and go to bed. I don't have to work now."

I looked at the men again. Most of them were hurrying; their faces carried yesterday's fatigue and last year's. Now and then I saw a man who looked as if he could work the turn and then box a little in the evening for exercise. There were a few men like that. The rest made me think strongly of a man holding himself from falling over a cliff, with fingers that paralyzed slowly.

I stepped on a stone and felt the place on my heel where the limestone and sweat had worked together, to make a burn.

I'd be hurrying in at 5.00 o'clock that day, and they'd be going home. It was now 7.20. That would be nine and a half hours hence. I had to eat twice, and buy a pair of gloves, and sew up my shirt, and get sleep before then. I lived twenty minutes from the mill. If I walk home, as fast as I can drive my legs and bolt breakfast, seven hours is all I can work in before 3.30. I'll have to get up then to get time for dinner, fixing up my shirt, and the walk to the mill.

I wonder how long this night-shift of gray-faced men, with different sized dinner buckets, will be moving out toward the green gate, and the day-shift coming in at the green gate — how many years?

The car up from the nail mill stopped just before it dove under the railroad bridge.

"I'm in luck."

I suddenly had a vision of how the New York subway looked: its crush, its noise, its overdressed Jews, its speed, its subway smell. I looked around inside the clattering trolley-car. Nobody was talking. The car was filled for the most part with Slavs, a few Italians, and some negroes from the nail mill. Everyone, except two old men of unknown age, was under thirty-five. They held their buckets on their laps, or put them on the floor between their legs. Six or eight were asleep. The rest sat quiet, with legs and neck loose, with their eyes open, steady, dull, fixed upon nothing at all.

IV

EVERYDAY LIFE

I came into the mill five minutes late one morning, and went to the green check-house at the gate, to pick 1611, the numerical me, from the hook. A stumpy man in a chair looked up and said: "What number?"

I gave it. "An easy way to lose forty-three cents," I thought, feeling a little sore at the stumpy man, and going out through the door slowly.

I increased my step along the road to the open-hearth, and reached my locker just as the Pole who shared it was leaving.

"Goddam gloves!" he was saying. "Pay thirty-five cents — three days — goddam it — all gone — too much. What you think?"

"I think the leather ones at fifty cents last better," I said.

He made a guttural noise, signifying disgust, and left.

I opened the locker, and disentangled my working-clothes, still damp from the last shift, from the Pole's. I removed all my "good" clothes, and stood for a minute naked and comfortable. The thermometer had registered 95° when I got up, at 4.00.

For the past few days I had been demoted to the pit; there had been no jobs open on the floor. As I took up my gloves and smoked glasses, I wondered how I could get back to furnace work.

Pete was moving with his lurching short steps past Six.

"How about helping to-day on the floor?" I said.

He snapped back quickly in his blurred voice, "You work th' pit, tell y' — goddam quick, want y' on the floor."

I looked back at him, swore to myself, and went slowly down the pit stairs.

I could n't find the gang at first, but later found half of them: Peter the Russian, the short Wop, the Aristocrat, and a couple more, all under furnace Eight, cleaning out cinder. The Aristocrat was trying to get the craneman to bring up one of the long boxes with curved bottoms for slag. The craneman was damning him. There was one too many at the job — four is enough to clean cinder — so I threw a bit of slag at Peter (for old time's sake) and passed on.

I met Al, and said, "Where are they working?"

"Clean up the pipes," he said.

The Croat, Marco, Joe, and Fritz were at Number 6, with forks. You see, the pipes run up the ladle's side and release a stopper for pouring the steel. They are covered with fire clay, which is destroyed after one or two ladlings and has to be knocked off and replaced. We loosened the clay with sledges,

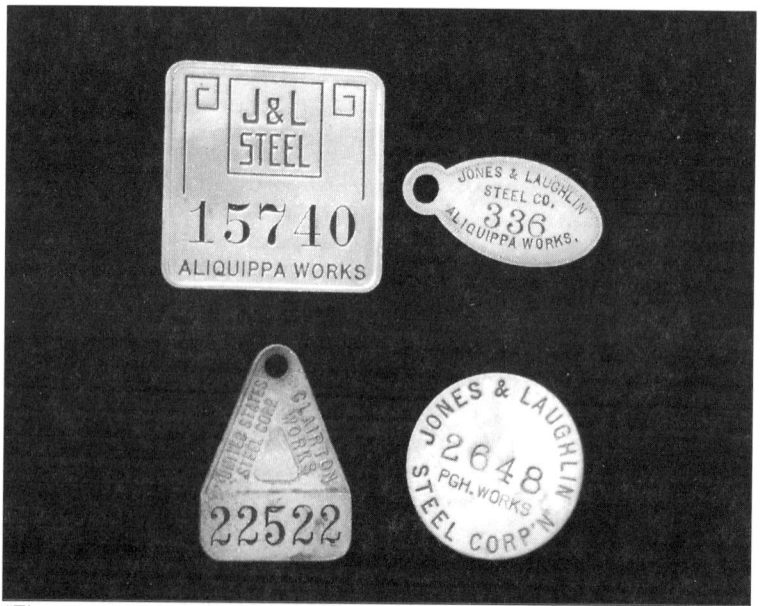

"The numerical me": Various company employee badges, including the Aliquippa Works.

and Marco watered down the pipes with a hose, to cool them. They were moderately warm when Fritz and I started piling them on the truck. Once or twice the pipe touched Fritz's hand through a hole in his glove, and he yowled, and then laughed. Once or twice I yowled and laughed also.

When we piled near the top, we swung in unison, and tossed the pipe into the air. It was like piling wood.

I caught a torn piece of my pants on a sharp bit of slag while carrying two pipes, and acquired a rip halfway from pocket to knee. Marco had a safety pin for me at once; he kept emergency ones in his shirt-front.

We finished the job in half an hour, and pushed the truck till it came under jurisdiction of a crane. Marco fixed the hooks rather officiously, pushing Fritz and me aside. There is, I suppose, more snobbishness induced by the manner of crane-hooking than in any other pit function. The crane swung the pipes on holders and dropped them in front of the blacksmith shop. We carried them into the shop, Marco and I working together. Inside there were half a dozen small forges, some benches, and a drop hammer. It was the place where ladles and spoons were repaired. The blacksmiths and helpers gave us friendly, but condescending glances.

As we walked back, we saw the crane swing a ladle from the moulds into which it had been pouring toward the dumping pit in front of Five. When the giant bucket approached, the chain hooked to the bottom lifted slowly, and dregs half-steel, half-ash, rolled out into the dump. After a little cooling, we would clean up there. With the chain released, the bucket righted itself with a shuddering clank, and swayed in the air scattering bits of slag and burnt fire clay.

A little later, we did a three-hour job on those dregs. We loosened the slag with picks first, and then lifted forkfuls and shovelfuls into the crane-carried boxes. A good deal of scrap was in the lot, probably the makings of half a ton of steel. This, of course, went into a separate box. I hooked up a couple of big scrap-hunks, weighing perhaps 500 pounds each, and took some sport out of it. That is one small matter, at least, where a grain of judgment and ingenuity has place. A badly hooked scrap-hunk may fall and break a neck, or simply tumble and waste everybody's time. Loosening up with the pick, too, demands a slight knack and smacks faintly of the miner's skill. We had to go down into a pit, where there was heated slag on all sides, using boards to save scorching our shoe leather. In turning up fractures eight or ten inches thick, there would be an inner four inches still red-hot.

At eleven o'clock, I was working at a fair pace, flinging moderately husky forkfuls over a ten-foot space into the box, when Marco looked up.

"Hey," he called.

I glanced at him for a moment. He was smiling. "Rest yourself," he said; "we work hard when de big bosses come."

During the next fifty shovelfuls, the remark went the rounds of my head, trying to get condemned. My memory threw up articles in the "Quarterly Journal of Economics," with inefficiency and the "labor-slackers," and "moral irresponsibility of the worker on the job," and so forth, in them. A couple of sermons and a vista of editorial denunciations of the laboring man who is no longer willing to do "an honest day's work for an honest day's pay," seemed to bring additional pressure for righteous indignation. I asked the following questions of myself, one for every two forkfuls: —

"Is n't it morally a bad thing to soldier, anyway?"

"Is Marco a moral enormity?"

"Do business men soldier?"

Is n't 'Get to hell out of here if you don't want to work' the answer? Or has the twelve-hour day something to do with it?"

EVERYDAY LIFE 61

"Can these five or six thousand unskilled workmen take any interest in their work, or must they go at it with a consciousness similar to that of the slaves who put up the Pyramids?"

I had to use the pick at this point, which broke up the inquiry, and I left the questions unanswered.

I saw wheelbarrowing ahead for the afternoon, and corralled the only one properly balanced, when I started work at 1.00 p.m., keeping it near me during a scrap-picking hour, until the job should break. At 2.15, it did. Al said: "Get over and clean out under Seven. If we can ever get this goddam stuff cleaned out —" That was an optimism of Al's.

One of the new men and I worked together all afternoon: pick at the slag, shovel, wheelbarrow, dump in the box, hook up to crane. Start over. There was a lot of dolomite and old fine cinder, very dusty, but not hot. This change in discomfort furnished a sensation almost pleasurable. I found out that everyone hid his shovel at the end of the shift, beside piles of brick in the cellar of the mill, under dark stairways, and so forth. I had n't yet acquired one, but used mostly a fork, which is n't so personal an instrument, and of which there seemed to be a common supply. I felt keen to "acquire" though.

After supper, I wrote in my I diary and thought a bit before going to bed. There 's a genuine technique of the shovel, the pick, and especially of the wheelbarrow, I thought. That damn plank from the ground to the cinder-box! It takes all I can muster to teeter the wheelbarrow up, dump without losing the thing quite, and bring it down backward without barking my shins. There 's a bit of technique, too, in pairing off properly for a job, selecting your lick of work promptly and not getting left jobless to the eyes of the boss, capturing your shovel and hiding it at the end of the turn, keeping the good will of the men you're with on team-work, distinguishing scrap from cinder and putting them into the proper boxes, not digging for slag too deeply in the pit floor, and so forth and so on.

I wonder if I shall learn Serbian, or Russian, or Hungarian? There seems to be a Slavic polyglot that any one of a half-dozen nationalities understands. That word, "Tchekai! — Watch out!" — even the Americans use it. It's a word that is crying in your ears all night. Watch out for the crane that is taking a ladle of hot metal over your head, or a load of scrap, or a bundle of pipes; watch out for the hot cinder coming down the hole from the furnace-doors; watch out for "me" while I get this wheelbarrow by; and "Heow! Tchekai!" for the trainload of hot ingots that passes your shoulder.

I set my alarm for five o'clock, and got into bed with the goodnight thought of "The devil with Pete Grayson! I'll get on that furnace!"

Another day went by, hewing cinders in the pit. I tried to figure to myself persuasive or threatening things I could say to the melters, to let me work on the floor. A shrewd-looking little man with moustachios worked near me.

"Did you ever work on the floor?" I asked.

"Oh, yes," he said, "too much hot; to hell with the money!"

They pay you two cents more an hour on the floor. At twenty minutes to five I went upstairs to my locker. Dick Reber, senior melter, stopped me.

"Need a man to-night; want to work?" he said; "always short, you know, on this — — long turn."

"Sure," I said.

That was one way to get promoted, I thought, and wondered how I'd stand fourteen more hours on top of the ten I had had.

"Beat it," yelled the melter.

Jack and I got our flat manganese shovels, and went on the run to the gallery. We were tapping at last. This furnaceful had cooked twenty-two hours. Nick was kneeling on water-soaked bagging, on the edge of the hot spout. He dug out the mud in the tap-hole with a pointed rod and sputtered oaths at the heat. Every few minutes the spout would burn through the bagging to his knees. He would get up, refold the bagging, and kneel again to the job.

Finally the metal gurgled out, a small stream the size of two fingers. Nick dodged back, and it swelled to a six-inch torrent.

"Heow, crane!"

Pete Grayson had come out, and was bawling something very urgently at the pit crane. The ladle swung closer; we could feel the increased wave of heat.

He looked over at us and held up two fingers. That meant both piles of manganese that lay on the gallery next the crane were to be shoveled in — double time for us, in the heat.

"Heow!" yelled the melter.

Jack and I leaped forward to the manganese, and our shovels scraped on the iron gallery. I saw Jack slapping his head to put out a little fire that had started on the handkerchief wound round his neck. I slapped a few sparks that stung my right leg. We finished half the pile.

There was something queer about this heat. The soles of my feet — why in hell should the gallery burn so! There was a blazing gas in the air — my nostrils seemed to flame as they took it in. This was different from most manganese shoveling. My face glowed all over in single concentrated pain. What was it? I saw Jack shoveling wildly in the middle of that second pile. We finished it in a panic.

"What was the matter with that damn ladle?" I asked, as we got our breath in the opening between the furnaces.

"Spout had a goddam hole in the middle," he said; "ladle underneath, see?"

I did. The fire-clay of the spout had given way, and a hole forming in the middle let the metal through. That made it necessary, in order to catch the steel, to bring the ladle close, till part of it was under the platform on which we worked. The heat and gas from the hot steel in the ladle had been warming the soles of our feet, and rising into our faces.

"Here's a funny thing," I said, looking down. One of the sparks which had struck my pants burned around, very neatly taking off the cuff and an inch or two of the pant-leg. The thing might have been done with a pair of shears.

I came out of the mill whistling and feeling pretty much "on the crest." I'd worked their damn "long turn," and stood it. It was n't so bad, all except that ladle that got under the manganese. I ate a huge breakfast, with a calm sense of virtue rewarded, and climbed into bed with a smile on my lips.

The alarm clock had been ringing several minutes before I realized what it was up to. I turned over to shut it off, and found needles running into all the muscles of my back. I struggled up on an elbow. I had a "hell of a head." The alarm was still going.

I fought myself out of bed and shut it off; stood up and tried to think. Pretty soon a thought came over me like an ache: it was "Fourteen hours!" That was beginning in fifty-five minutes — fourteen hours of back-walls, and hot ladles, and — Oh, hell! — I sat down again on the bed and prepared to lift my feet back in.

Then I got up, and washed fiercely, threw on my clothes, and went downstairs, and out into the afternoon sun.

Down by the restaurant, I met the third-helper on Eight.

"Long turn would n't be so bad, if there were n't no next day," he said, with a sort of smile.

In the mill was a gang of malignant men; things all went wrong; everybody was angry and tired; their nerves made mistakes for them.

"I only wish it were next Sunday!" I said to someone.

"There are n't any goddam Sundays in this place," he returned. "Twenty-four hours off between two working days ain't Sunday."

I thought that over. The company says they give you one day off every two weeks. But it's not like a day off anywhere else. It's twenty-four hours sandwiched between two work-days. You finish your night-week at 7.00 Sunday morning, having just done a week of one twenty-four hour shift, and six fourteens. You've got all the time from then till the next morning! Hurrah! How will you use it? If you do the normal thing, — eat breakfast and go to bed for eight hours, — that brings you to 5.00 o'clock. Will you stay up all night? you've had your sleep. Yes, but there's a ten-hour turn coming at 7.00. You go to bed at 11.00, to sleep up for your turn. There's an evening out of it! Hurrah again! But who in hell does the normal thing? Either you go on a tear for twenty-four hours, — you only have it twice a month, — or you sleep the twenty-four, if the week's been a bad one. Or — and this is common in Bouton — you get sore at the system and stay away a week — if you can afford it.

"Hey, you, get me a jigger, quick. Ten thou'."

"All right," I said, and shut off my mind for the day.

I usually had bad words and bad looks from "Shorty." Jack calls him "that dirty Wop." Late one afternoon he produced a knife and fingered it suggestively while he talked. So I always watched him with all the eyes I had.

One day we had shoveled in manganese together over a hot ladle, and I noticed that he was in a bad mood. We finished and leaned against the rail.

"Six days more," he said very quietly.

I looked up, surprised at his voice.

"What do you mean?"

"Six days more, this week, me quit this goddam job."

"What's the matter?"

"Oh, —— me lose thirteen pound this job, what the hell!"

"What job will you get now?"

"I don't know, I don't know; any damn job better than this," he said very bitterly.

Having adopted the quitting idea, these six days were too much to endure. A little later, Jock was ready to make frontwall. He saw Shorty and said, "Get me that hook and spoon."

Shorty stood and looked at Jock, with the utmost malignity in his face, and said finally, "Get your goddam hook and spoon yourself."

Jock was greatly surprised, and returned, "Who the hell are you?"

Shorty snapped instantly, "Who the hell are you?"

And then he was fired.

This is the second "quitting mad" I've seen. The feeling seems to be something like the irrepressible desire that gets piled up sometimes in the ranks of the army to "tell 'em to go to hell" and take the consequences. It's the result of accumulated poisons of overfatigue, long hours, overwrought nerves, "the military discipline of the mills."

The practical advantage of being "given the hook" is that you can draw your pay immediately; whereas, if you simply leave, you have to wait for the end of the two weeks' period.

I ate my dinner at the Greek's.

"Make me some tea that's hot, George. This was n't. Oh, and a double bowl of shredded; I've got a hole to fill up."

George kept the best of the four Greek restaurants. It had a certain variety. It splurged into potato salad, and a few other kinds, and went into omelets that were very acceptable. The others confined themselves to fried things, with a few cereals and skimmed milk. I looked up from my shredded wheat. George was wiping up a rill of gravy and milk from the porcelain table, and a man was getting ready to sit down opposite. It was Herb, the pit craneman.

"Always feed here?" he asked.

"Yes," I said, "best place in town, is n't it?"

He nodded.

"How big is Bouton? how many people has it?" I asked.

He grinned slowly, and put his elbows on the table. He was a Pennsylvania Dutchman, with worry settling over good nature in a square face.

"Twenty thousand," he said.

"It seems small for twenty thousand," I returned; "like a little village. There's really only one store, is n't there — the company store — where they keep anything? Only one empty newspaper, no theatre, unless you count that one-story movie place, no enterprise — "

"A one-man town," he said, quickly. "Nearly every house in town is owned by Mr. Burnham. Now look here, suppose a man works like hell to fix things up, to work around and get a pretty damn good garden, puts a lot of money into making his house right. Suppose he does, and then gets into a scrap with his boss. What can he do? The company owns his house, the company owns every other damn thing in town. He's got to beat it — all his work shot to hell. That's why nobody does anything. Hey, ham and — Where you workin' now? Ain't seen yer in the pit."

"I'm on the floor, helpin' on Number 7."

"Att-a-boy!"

At last, Saturday night. Everyone felt Sunday coming, with twenty-four hours of drunkenness or sleep alluringly ahead.

The other shift had tapped the furnace at three o'clock. We might not tap again, and that was nice to think about. A front-wall and a hot back-wall we went through as if it were better fun than billiards.

"Look out for me, I've got the de'il in me," from Jock, Scotch First on Number 8. I looked up, and the crazy fool had a spoon — they weigh over a hundred — between his legs, dragging it like a kid with a broomstick. As it bounced on the broken brick floor, he yelled like a man after a Hun.

"Who's the maun amang ye, can lick a Scotchman?" he cried, dropping the spoon to the floor.

"Is this the best stuff you can show on Number 8?" said Fred slowly. He dived for Jock's waist, and drew it to him, though the Scotchman tried to break his grip with one of his hands and with the other thrust off his opponent's face. When Fred had him tight, he caught one of Jock's straying arms, bent it slowly behind his back, and contrived a hammerlock.

"You're no gentlemen," — in pain; "you're interruptin' my work."

Fred relaxed, and Jock jumped away.

"Come over to a good furnace, goddam it, and fight it out!" he yelled, from a distance that protected his words.

The charging-machine, in its perpetual machine-tremolo, shook past and stopped. George slid down from his seat and came over to Number 8's gang.

"Well, Fred, how in hell's the world usin' yer?"

"Ask me that to-morrow."

"Well, guys, good night; I'm dead for forty minutes."

He picked up a board some six feet long and about six inches in width. He laid himself carefully on it, and was sleeping inside of a minute.

I looked at him enviously for a few minutes. Suddenly it occurred to me that the board lay over a slit in the floor. It was the opening through which the pipes that attach to the gas-valve rise and fall. When gas is shifted from one end of the furnace to the other, the pipes emerge through the slit to a height several feet from the floor. Finally, Fred made the same discovery, and a broad smile spread over his face. He continued to watch George, his grin deepening. At last he turned to the second-helper.

"Throw her over," he said.

Nick threw the switch. Slowly and easily the valve-pipes rose, lifting George and the head of his bed into the air, perilously. An immense and ill-controlled shout swelled up and got ready to burst inside the witnesses. George slept on, and the bed passed forty-five degrees. In another second it rolled off the side of the pipes, and George, scared, half-asleep, and much crumpled, rolled over on the furnace floor. It was several seconds before he recovered profanity.

The pure joy of that event spread itself over the entire shift.

When the light from the melting scrap-iron inside the furnace shot back, it lit up the hills and valleys in Nick's face. I noticed how sharp the slope was from his cheek-bones to the pit of his cheeks, and the round holes in which his eyes were a pool at the bottom. His lips moved off his white teeth, and twisted themselves, as a man's do with effort. He looked as if he were smiling. I picked up my shovel, and shoved it into the dolomite pile, with a slight pressure of knee against right forearm that eases your back. The thermometer in the shade outside was 95°. I wondered vaguely how much it was where Nick stood, with the doors open in his face.

We walked back together after the front-wall to the trough of water.

"Not bad when you get good furnace, good first-helper," he said. "Fred good boy, but furnace no good. A man got to watch himself on this job," he went on bitterly; "he pull himself to pieces."

"I can't manage quite enough sleep," I said, wondering if that was the remark of a tenderfoot.

"Sometime — maybe one day a month — I feel all right, good, no sleepy," he went on. "Daytime work, ten hour, all right, feel good; fourteen hour always too much tired. Sometime, goddam, I go home, I go to bed, throw myself down this way." He threw both arms backward and to the side in a gesture of desperate exhaustion, allowing his head to fall back at the same time. "Goddam, think I no work no more. No day nuff sleep for work," he concluded.

Later on in the day, I saw Jimmy let the charge-up man, George, take the spoon and make front-wall. The heat "got his goat." "I lose about ten or fifteen pounds every summer," he said, "but I get it back in the winter. My wife is after me the whole time to leave this game. I tell her every year I will. Better quit this business, buddy, while you're young, before you get stuck like me."

I walked home with Stanley, the Pole. He always called me Joe, the generic name for non-Hunky helpers.

"Say, Joe," he said, as we came under the railroad bridge, "what's your name right?"

"Charlie," I answered. "By the way, where have you been?"

"Drunk, Charlie," he answered, smiling cheerfully.

"Ever since I saw you in the pit?"

"Three week," he stated, with satisfaction; "beer, whiskey, everyt'ing. What the hell, work all time goddam job, what the hell?"

V

WORKING THE TWENTY-FOUR HOUR SHIFT

7 A. M. *Sunday*

I TRIED to get a lot of sleep last night for handling the long turn; managed about nine hours. When I came to the locker, Stanley was there, dressed, cleaning his smoked glasses.

"How much sleep last night?" I asked.

"Oh, six, seven hour," said Stanley.

"You're a damned fool," I said; "this is the long turn."

"I know, I know," he returned, "I have t'ing to do. No have time sleep."

I looked at him. He had a big frame, but his limbs were hung on it, like clothes on hooks. His face was a gray pallor, sharply caving in under the cheek-bones. His eyes were very dull, and steady. I'd noticed those eyes of his before, and never could decide whether they showed a kind of sullen defiance, or resignation, or were just extraordinarily tired.

"Two month more," he said.

"Two month more what?"

"Two month more this work every Sunday — goddam work all day like hell, all night like hell. Pretty soon go back to good job."

I knew what he meant now. He had told me weeks before, when we had hewed cinders together in the pit, how he was a rougher in a Pittsburgh mill. Worked only twelve hours a day and no Sundays.

"No more goddam long turn," he concluded; "work of rougher slack now, all right October."

He moved off slowly, with no spring in this step, and no energy expended beyond what was absolutely necessary to move him.

I walked out on the floor to look at the clock. The night gang on every furnace was washing up, very cheerfully and with an extraordinary thoroughness. They were slicking up for the once a fortnight twenty-four-hour party. Nearly everyone drank through his day off, or raised hell in some extraordinary manner. It was too precious and rare to spend in less violent reaction to the two weeks' fatigue. I looked at them and tried not to be envious. The first-helper on Seven was taking a last look through the peepholes as he put on his collar. A great Slavic hulk on Number 5 was brushing his clothes with unheard of violence.

Dick Reber passed by. He saw me leaning against a girder buttoning my shirt.

"Front-wall, Number 5, you!" he bawled.

I was sore at myself for having been seen standing about doing nothing. But I was sore at Dick also, unreasonably. I went back to my locker, got my gloves, and went to Number 5. I began filling the spoon, with the help of "Marty," the Wop. He glared at me, and interfered with my shovel twice when we went together to the dolomite pile. Marty had made enemies widely on the furnaces because of a loud mouth, and an officiousness that sat ridiculously on his stature and his ignorance of steel-making.

I was glad when the front-wall was done. I took the hook down, and went over to the fountain in back of Five, cooled my head, neck, and arms, and went over to Seven, without taking a swallow. I had decided to have only two drinks of water in the half-day.

Dick Reber saw me coming up and, I think in punishment for loafing, said, "Clean up under there. I want you to clean all that filth out, all of it, from behind that girder."

It was near the locker and under the flooring, in a sort of shelf, where lime, dolomite, dirt, old gloves, shoes, filth of all

THE TWENTY-FOUR HOUR SHIFT

sorts had accumulated. I cleaned it out with a broom and a stick. It took me half an hour.

"All right," said the first-helper: "now get me ten thousand."

So I went off to the Bessemer, rather glad of the walk. I climbed the stairs to the pouring platform, and watched the recorder, who had left his book, operate the levers. The shifting engine backed a ladle under, and slowly the huge pig-iron mixer, bubbling and shooting out a tide of sparks, dipped and allowed about 20,000 pounds to drop into the ladle.

"Ten thou' for Seven," I said.

In another five minutes, the engine brought up a ladle for my ten thousand, and the boy dipped it out for me with the miraculous levers.

Charging iron into a fixed open hearth. Homestead (PA) Works, Open Hearth Shop #5, Sept. 1953, United States Steel Corporation. Photo: Archives of Industrial Society, Wm. J. Gaughan Collection, University of Pittsburgh.

"All right," I said; and ran down the stairs fast enough to catch a ride back past the furnaces, on the step of the locomotive.

The second-helper grabbed the big hook which came down slowly on a chain from the crane, and stuck it into the bottom of the ladle. As the chain lifted, the ladle tipped and poured the ten thousand pounds with a hiss. But the craneman was careless, which is n't usual. Fred kept saying, "Whoop, whoop!" but he went right on spilling for quite a spell before he recovered control.

"Dolomite," said the first-helper to me, after the "jigger" was poured.

I went to a box full of the white gravel, at the end of the mill, and yelled at Herb, the craneman. A box of dolomite is about eight feet square and three high. This one was perched on top of a dolomite pile, ten feet off the ground. I struggled up on top, and took the hooks Herb gave me from the crane, — eight-inch hooks, — and put them into the corners of the box, using both hands. Then I slid down, and the box rose and swung over my head.

Herb settled it neatly on our own little dolomite pile in front of Seven. I slipped out the front hooks, and the back ones lifted and dumped the load, with a soft swish, nearly on the low part of the old pile.

There was a little time to sit down after this — perhaps ten minutes. I smoked a Camel, which had spent the last shift in my shirt pocket. It was a melancholy Camel, and tended to twist up in my nose, but it tasted sweet. I sat on Seven's bench, and watched Fred take his rod and move aside the shutters of the peepholes, to give final looks at the furnace. She must be nearly ready. He looked back at me, and I knew that meant "test."

I grabbed tongs, lying spread out by the anvil, clamped hold of the mould, and ran with them to about ten feet from number two door of the furnace. Fred had the test-spoon lifted and shoved into the door; he moved it around in the molten steel, and brought it out full, straining his body tense to hold it level and not lose the test. I shifted the mould a little on the ground, and closed my hands as tight as I could on the tongs, so the mould would n't slip and turn. He poured easily and neatly, just filling the mould, and flung the spoon violently on the floor, to shake off the crusting steel on the handle.

I ran with mould and tongs to the water-trough in front of Eight, and plunged it in, the steam coming up in a small cloud. I brought it out and held it on the anvil, end-wise, with the tongs, while Nick flattened in the top slightly on both edges, to make it break easily. Nick broke the ingot in two blows, and Fred and the melter consulted over the fragments.

To tap some fixed furnaces, a large bar was inserted through the charging side of the furnace. It was poked through the tap hole with a mighty heave of the workers, as shown here. Homestead (PA) Works, Open Hearth Shop #5, July 1951, United States Steel Corporation. Photo: Archives of Industrial Society, Wm. J. Gaughan Collection, University of Pittsburgh.

"All right," said Dick.

We were about to tap. I went after my flat manganese shovel, but it was gone from the locker. Some dog-gone helper has nailed it. I took out an ordinary flat shovel.

Sometimes dynamite was used to open the tap hole. This was called "jet tapping." Here we see the surge of metal into the ladle at the beginning of the tap. An additive box is suspended above the ladle; this replaced men like Walker who had done this job previously.
Homestead (PA) Works, Open Hearth Shop #5, Furnace #65, Dec. 30 1948, United States Steel Corporation. Photo: Archives of Industrial Society, Wm. J. Gaughan Collection, University of Pittsburgh.

In back of the furnace Nick was already busy with a "picker," prodding away the stopping from the tap. He burned his hands once, swore, gave it up, went halfway along the platform away from the tap, returned, and went at it again. Finally, the steel escaped with its usual roar of flame and its usual splunch as it fell into the ladle.

I stepped back, and nearly into "Shorty," who had come to help shovel manganese. "Where you get shovel?" he said, with his eyes blazing, pointing to mine.

"Out of my locker," I said.

He started toward it, and I held it away from him.

"I tell you that goddam shovel mine —" he began; but Dick, from the other side of the spout, shouted at us how many piles to shovel, and Shorty shut up. We were to get in the first big pile and the next little one.

The ladle was beginning to fill. "Heow!" yelled Dick.

Shorty and I went forward and put in the manganese. It was hot, but I took too much interest in shoveling faster than Shorty, to care. Then came the second ladle, during which Shorty's handkerchief caught on fire, and made him sputter a lot, and rid himself of some profanity in Anglo-Italian.

I went to that trough by Eight afterward, to wash off the soot and cinder, and put my head under water, straight down. I knew back-wall was coming, and sat down a minute, wondering, rather vaguely, how I was going to feel at six or seven the next morning.

Back-wall came. I had bad luck with it, trying too hard. It was too hot for one thing. There are times when a back-wall will be so cool you can hesitate a long second, as you fling your shovelful, and make sure of your aim; at others, your face scorches when you first swing back, and you let the stuff off any fashion, to get out of the heat. There's a third-helper on Five, I'm glad to say, who is worse than I. They put him out of the line this time; he was just throwing into the bottom of the furnace.

Everyone develops an individual technique. Jimmy's is bending his knees, and getting his shovel so low that it looks like scooping off the floor. Fred's is graceful, with a smart snap at the end.

Then front-wall. I start in search of a spoon and a hook. It's not easy to get one to suit the taste of my first-helper. There's one that looks twenty feet, — I have n't any technical figures on spoons, — but it's too long, I know, for Fred. There's a spoon

Slag running onto the pit floor during a tap. Homestead (PA) Works, Open Hearth Shop #5, Sept. 1953, United States Steel Corporation. Photo: Archives of Industrial Society, Wm. J. Gaughan Collection, University of Pittsburgh.

three feet shorter, just right. Hell — with two inches melted off the end! I pick a short one in good repair, — he can use the thing or get his own, — and drag it to Seven, giving the scoop a ride on the railroad track, to ease the weight. Fred has put a hook over number one door; so I hurry, and lift the spoon handle with gloved hands to slip it on the hook. If it's not done quickly, you'll get a burn; you're an arm's length from molten steel, and no door between. I get it on, and pick up a shovel.

Front-wall can be very easy, — you can nearly enjoy it, like any of the jobs, — if the furnace is cool, and there's a breeze blowing down the open spaces of the mill. And, too, if the spoon hangs right in the hook, and the first-helper turns it a little for you, then you can stand off, six feet from the flame, and toss your gravel straight into the spoon's scoop. You hardly go to the water fountain to cool your head when the stunt's over. On number one the hook hung wrong, the spoon would n't turn in it, and you had to hug close, and pour, not toss. I tried a toss on my second shovel, and half of it skated on the floor.

"Get it on the spoon, goddam you!" from Nick.

So I did.

After that, we sat around for twenty minutes. Fred looked at the furnace once or twice, and changed the gas. Several gathered in front of Seven — Jock, Dick, the melter, Fred, and Nick.

"Do you know what my next job's going to be?" said Fred.

The others looked up.

"In a bank."

"Nine to five," said Dick. "Huh! gentlemen's hours."

"Saturday afternoons, and Sundays," said Fred.

The other faces glowed and said nothing.

"This would n't be so bad if there were Sundays," said Fred.

"I'll tell you, there'll come a time," broke in the melter,

"when Gary and all the other big fellers will have to work it themselves — no one else will."

"Now in the old country, a man can have a bit of fun," said the Scotchman. "Picnics, a little singin' and drinkin', — and the like. What can a man do here? We work eight hours in Scotland. They work eight hours in France, in Italy, in Germany — all the steel mills work eight hours, except in this bloody free country."

The melter broke in again. "It's the dollar they're after — the sucking dollar. They say they're going to cut out the long turn. I

Furnace workers crossed all ethnic and racial boundaries. Here two black workers and a white worker with blue glasses in his hand are about to pour a test. Homestead (PA) Works, Open Hearth Shop #5, ca. 1954, United States Steel Corporation. Photo: Archives of Industrial Society, Wm. J. Gaughan Collection, University of Pittsburgh.

heard they were going to cut out the long turn when I went to work in the mill, as a kid. I'm workin' it, ain't I? Christ!"

I left, to shovel in fluor spar with Fred.

When we finished, Fred said: "You better get your lunch now, if you want it. Then help Nick on the spout."

I ate in the mill restaurant. My order was roast beef, which included mashed potato, peas, and a cup of coffee — for thirty-five cents. Then I had apple pie and a glass of milk. The waiters are a fresh Jew, named Beck, and a short, fat Irish boy, called Pop. There is a counter, no tables; the food is clean.

I went back to help Nick on the spout, and found him already back, on the gallery with a wheelbarrow of mud. He looked up gloomily and said: "One more."

I dumped the wheelbarrow, and went after more, bounced it over tracks and a hose, and up and down a little board runway to where the mud-box stands. After filling up, I went back slowly, dangerously, swayingly, over bits of dolomite and coal, navigated the corner of the gallery by a hair's tolerance, and dropped the handles of the wheelbarrow by Nick with relief. It's bad on my back, that's it. I'd rather do two back-walls, and tap three times in high heat, than wheel these exacting loads of mud.

Nick knelt on the other side of the spout, and I gave him the mud with my shovel to repair the holes and broken places of the spout, which the last flow of molten steel had carried away. When he finished the big holes, I gave him small gobs of mud, dipping my hands in a bucket of water between each two, to keep the stuff from sticking. A wave of weakening heat rose constantly from the spout, still hot from the last flow. I prayed God Nick would hurry. He made a smooth neat surface on the whole seven-feet of spout, rounding the edges with his hands.

When I came back from the spout, Fred was in front of the furnace, blue glasses on his nose, inspecting the brew. He put his glasses back on his cap, glanced at me, and pointed to a pile of dolomite and slag which had been growing in front of Number 3 door.

"All right," I said, and picked up a shovel from the dolomite pile. For a couple of minutes, I shoveled the stuff down the slag hole, and remembered vividly the bygone pit-days. Then I would have been cleaning up around the buggy. For a minute I felt vastly superior to pit people. I earned two cents more an hour, and threw down a hole the dolomite and dirt they cleared away.

I began to feel a little tired in back and legs, and repeated Fred's formula on how to get away with a long turn: "Take it

As with Fred in the text, this first helper "put his glasses back on his cap." The furnace door is open and a small amount of slag is running out the door. Homestead (PA) Works, Open Hearth Shop #5, Mar. 1949, United States Steel Corporation. Photo: Archives of Industrial Society, Wm. J. Gaughan Collection, University of Pittsburgh.

like any other day to five o'clock. Then work for midnight. Anyone can stand it from midnight to morning." I did a front-wall on that basis.

"Watch those buggies!"

I ran over to the furnace and glanced down the slag hole, yelling back, "Half full." Then Fred went to an electric switch and the whole furnace tilted till the hot running slag flowed over at the doors, and dripped into the buggy-car beneath, in the pit. I held my hand up as one of them filled, and Fred caught the pitching furnace with the switch, and stopped the flow of slag.

4 P. M. *Sunday*

Number 8 furnace tapped, and I shoveled manganese into the ladle with that man from Akron, who is new, and who, I noticed, burned his fingers in the same way I did on my first day. Then back-wall and front-wall, and Jock saying all the while, "It's a third gone, lads."

5 P. M. *Sunday*

I felt much more tired after this first ten hours than later; it was the limp fatigue that comes from too much heat. I ate fried eggs and a glass of milk, and then my appetite took a start and I ordered cold lamb and vegetables. When I finished, I went back into the mill to my locker, and took out a cigarette. I sat on a pile of pipes against a main girder, intending to smoke; the cigarette went out, and I slept a half hour.

Things were going first-rate from six to nine. Jigger, clean up scrap, front-wall Number 6, front-wall Number 8. I could n't distinguish between this and any other night shift; the food must have acted for sleep. But after nine the hours dragged. From 9.20 to 10.00 was a couple of hours.

In the middle of a front-wall, I saw the efficiency man, Mr. Lever, come through and stare at the furnace, walk around a little, and stare profoundly at the furnace. Mr. Lever was pointed in two places, I noticed for the first time. He had a pointed stomach, and his face worked into a point at his nose. I noticed carefully that he had a receding chin and a receding forehead. As he watched us scoop the dolomite, drag up to the spoon, dump, scoop up the dolomite, and do it over, for three quarters of an hour, I thought about him. I wanted to go up to him, and give him my shovel. I had to struggle against that impulse — to go up to him and give him my shovel.

The evening dragged. I fought myself, to keep from looking at the clock. I fought for several hours after ten o'clock, and then, when I thought dawn must be breaking, went up and found it ten minutes of eleven.

I did feel relieved at twelve, and went out to the restaurant, saying: "Hell, anyone can wait till morning."

Sometimes, when things are hurried, when tapping is near or a spout is to be fixed, you have to eat still drenched in sweat. But to-night I had time, and at quarter of twelve hung my shirt on the hot bricks at the side of the furnace, and stood near the doors in the heat, to dry my back and legs. I then washed soot and dolomite dust from ears and neck, and dipped my left arm, which was burned, in cold water. At twelve I put on the dried shirt, and went to eat.

Half the men wash, half don't. There were a number of open-hearth helpers in the restaurant, with black hands and faces, two eating soup, two with their arms on the counter. Their faces lacked any expression beyond a sullen fatigue; but their eyes roved, following Beck about. Lefflin had his arms on the counter and his face on them.

I ate ham and eggs, which included coffee, fried potatoes, two slices of bread, and a glass of milk.

Walking back to the furnaces was an effort of will. I climbed the embankment to the tracks very slowly, the stones and gravel loosening and tumbling down-hill at each step. I tried hard to concentrate on a calculation of the probable number of front-walls to come. Then I wondered if it would n't pay to cut out breakfast in the morning, and get nine hours of sleep instead of eight and a quarter. Friselli came up the bank behind me. He is third on Number 6.

"Well" I said, "make lots of money to-night."

"What's the good money, kill yourself?" he said, and went past me along the tracks.

Number 8 was preparing to make front-wall. I felt weary, and full of ham and eggs, and very desirous of sitting down right there on the floor. But Jock, the first-helper on Eight, said, "Oh, Walker!" when he saw me, and we began.

Through that front-wall Jock was tiring. He worked in little spurts. For "half a door" he would sing, and goad us on in half-Scotch, and for the next half he would be silent, and wipe his face with his sleeve. After that door, he came up to us and said with profound conviction, "It's a lang turn, it's a lang turn."

When we finished, Jock lay down on a bench.
It's a part of a third-helper's duties to keep five or six bags of fine anthracite coal on the little gallery back of the furnace, near the spout. I went after that little job now. Fifty pounds of coal in a thick paper bag is n't much to carry, till you get doing it a couple of days running.

I sat on the seat where the Wop stays who works the furnace-doors; they call him the "pull up." That had some sacks and a cushion, and was broad, with a girder for back. I fell asleep.

Something twisting and pinching my foot woke me up. It was the first-helper. "Fifteen thousand, quick!" he said. I got up with a jerk, feeling not so sleepy as I expected, but immeasurably stiff. I moved in a wobbly fashion down toward the Bessemer. I felt as if I were limping in four or five directions. Very vigorously and insistently I thought of one thing. I would look at the clock opposite Number 6 when I went by, and possibly, very probably, a whole pile of hours had been knocked off. Then I thought with a sting that we had not tapped, and it could n't be more than three o'clock. It was two!

"Fifteen thousand" I said to myself, "quick"; and climbed the iron stairs to the Bessemer platform.

When I came back, I walked beside the locomotive as it dragged the ladle and the fifteen thousand pounds of molten pig iron. Through closing eyes I watched the charging-machine thrust in the spout. That long finger lifted the clay thing from its resting-place on the big saw-horses between furnaces. Then, moving on the rails, the machine adjusted itself in front of number two door and shoved the spout in with a jar.

I stood lazily watching the pouring of the molten steel. Fred motioned slowly with his hands, with "Up a little, whoop!" as the stream flowed very cleanly into the spout and furnace. Then came the noise of lifting, that characteristic crane grind, with a rising inflection as it gained speed and moved off. "Pretty soon tapping, after tapping back-wall, front-wall, the spout, morning," I meditated.

"Well, how in hell are you?" It was Al, the pit boss.

"Fine!" I said as loudly as I could; and went and sat down at once. My chin hit my chest. I stopped thinking, but did n't go to sleep.

"Test!" yelled Fred.

We tested three times, and then tapped. There were two ladles, with four piles of manganese, to shovel in. A third-helper from Number 4, a short stocky Italian, shoveled with me. The ladle swung slightly closer to the gallery than usual, and sent up a bit more gas and sparks. We put out little fires on our clothes six or seven times. After the first ladle, the Italian put back the sheet iron over the red-hot spout, and after the second ladle, I put it on. We rested between ladles, in a little breeze that came through between furnaces.

"What you think of this job?" he asked.

"Pretty bad," I said, "but pretty good money."

He looked up, and the veins swelled on his forehead. His cheeks were inflamed, and his eyes showed the effects of the twenty hours of continuous labor.

"To hell with the money!" he said, with quiet passion: "no can live."

The words sank into my memory for all time.

The back-wall was, I think, no hotter than usual, but men's nerves made them mind things they would have smirked at the previous morning. The third-helper on Eight and Nick quarreled over a shovel, and Nick sulked till Fred went over and spoke to him. Once the third-helper got in Nick's way. "Get out, or I'll break your goddam neck!" And so on —

I felt outrageously sore at everyone present — not least, myself. After that back-wall all except Fred threw their shovels with violence on the floor, and went to the edge of the mill. They stood about in the little breeze that had come up there, in a state of fatigue and jangled nerves, looking out on a pale streak of morning just visible over freight cars and piles of scrap.

We made front-wall, and when it was over, I went to the bench by the locker and sat down, to try to forget about the spout. I had been forgetting about it for twenty minutes when Nick came up, and shook me, thinking I had fallen asleep.

"Mud," he said.

I got him mud.

Nick fixed up the spout amid an inclination to cursing in Serbian, and gave me commands in loud tones in the same language. I felt exceedingly indifferent to Nick and to the spout, and finished up in a state of enormous indifference to all things save the chance of sleep. Jack, the second-helper of Eight, was making tea, having dipped out some hot steel with a test-spoon, and set a tea-pot on it.

"Want some?" he said.

I nodded.

Watching him make it, and drinking the tea woke me up.

"What time is it?" I asked.

"Four-thirty," said he.

"Thanks for the tea."

Then the summoning signal for a third-helper rang out — a sledge-hammer pounding on sheet iron. They were "spooning up," that is, making front-wall, on Number 6. All through that stunt I was wide awake, quite refreshed, though with the sense, the conviction, that I had been in the mill, doing this sort of thing, for a week at the inside.

Coming back to Seven from that, I found Fred flat on his back, looking "all in." Jock came up for a drink of water, and looked over at me.

"You look to me," he remarked, "like the breaking up of a bad winter." He laughed.

5 A. M. *Monday*

The sun came into the mill, looking very pallid and sick beside the bright light from the metal. I watched the men on Eight make back-wall, and heard the sounds; I sat on the bench, my legs as loose as I could make them, my head forward, eyes just raised.

"Lower, lower, goddam you, lower!" came a desperate command to the "pull-up" man, to close the furnace doors.

"Get out —"

"One more —"

"Up, up goddam it! where are your ears?"

"Come on, men, last door."

"My shovel you son-of-a — !"

Now they were tapping on Number 6. The melter came out of his shanty; he had had a sleep since the last furnace tapped. He rubbed his eyes, and went out on the gallery. I could hear his "Heow." Four poor devils were standing in the flame, putting in manganese. Thank God, I don't shovel for Six.

"A jigger," from Fred.

"Sure."

When I went for it, the sores on the bottom of my feet hurt, so that I walked on the edges of my shoes. I was so delighted with the idea of its being six o'clock, with no back-walls ahead, that I almost took pleasure in that foot. I stopped in front of a fountain and put my right arm under the water.

The recorder in the Bessemer was asleep. He was a boy of twenty. I woke him up, and grinned in his face.

"Fifteen thou' for Number 7."

"You go to hell with your goddam Number 7!"

I grinned at him again, knew it was just the long turn, knew he'd give me that fifteen thousand pounds; went down stairs again —

Twenty minutes of seven. It's light. Nobody talks, but everyone dresses in a hurry. Everyone's face looks grave from fatigue — eyes dead. We leave at ten minutes of seven.

7 A. M. *Monday*

It's a problem — a damn problem — whether to walk fast and get home quick, or walk slow and sort of rest. I try to go fast, and have the sense of lifting my legs, not with the muscles, but with something else. I shake my head to get it clearer. One bowl of oatmeal. Coffee. "I feel all right." I get up and am conscious of walking home quietly and evenly, without any further worry about the difficulty of lifting my feet. "The long turns, they 're not so bad," I say out loud, and stumble the same second on the stairs. I get up, angry, and with my feet stinging with pain. Old thought comes back: "Only seven to eight hours sleep. Bed. Quick." I push into my room — the sun is all over my bed. Pull the curtain; shut out a little. Take off my shoes. It 's hard work trying to be careful about it, and it 's darn painful when I'm not careful. Sit on the bed, lift up my feet. Feet burning all over; wonder if I'll ever sleep. Sleep.

VI

BLAST-FURNACE APPRENTICESHIP

At the end of every shift, when I walked toward the green mill-gate just past the edge of the power house, I could look over toward the blast-furnaces. There were five of them, standing up like mammoth cigars some hundred feet in height. A maze of pipes, large as tunnels, twisted about them, and passed into great boilers, three or four of which arose between each two furnaces. These, I learned were "stoves" for heating the blast. I had had in mind for several days asking for a transfer to this interesting apparatus. There was less lifting of dead weight on the blast-furnace jobs than on the open-hearth. Besides, I wanted to see the beginning of the making of steel — the first transformation the ore catches, on its way toward becoming a steel rail, or a surgical instrument.

I went to see the blast-furnace superintendent, Mr. Beck, at his house on Superintendent's Hill.

"I'm working on the open-hearth," I said, "and want very much to get transferred to the blast-furnace. I intend to learn the steel business, and want to see the beginnings of things."

"How much education?" he asked.

"I graduated from college," I said, "Yale College." Would that complicate the thing, I wondered, or get in the way? I wanted badly to sit down for a talk, tell him the whole story — army, Washington, hopes and fears; I liked him a good deal. But he was in a hurry — perhaps that might come on a later day.

We talked a little. He said I ought to come into the office for a while and "learn to figure burdens." I replied that I wanted the experience of the outside, and a start at the bottom.

"All right," he said, "I'll put you outside. Come Monday morning."

On Monday morning I followed the cindered road inside the gate for three hundred yards, turned off across a railroad track, and passed a machine-shop. The concrete bases of the blast-furnaces rose before me. Somebody had just turned a wheel on the side of one of the boiler-like "stoves," and a deafening blare, like tons of steam getting away, broke on my eardrums. I asked where the office was.

"Through there."

Up some steps, over a concrete platform, past the blaring "stove," I went, to the other side of the furnaces, and found

Some of the houses on Superintendents' Hill; modern photo. Collection of Don Inman.

there a flat dirty building — the office. Inside was Mr. Beck, who turned me over at once to Adolph, the "stove-gang boss."

I was a little anxious over this introduction to things, and thought it might embarrass or prevent comradeships. But it did n't. No one knew, or if he did, ever gave it a thought. It may perhaps have accounted for Adolph's letting me keep my clothes in his shanty that night, and for considerable conversation he vouchsafed me on the first day. But my individuality passed quickly, very quickly; I became no more than a part of that rather dingy unit, the stove-gang.

While I was putting on my clothes in Adolph's sheet-iron shanty, he grinned and said "Last time, pretty dirty job, too, eh?"

"Yes," I said "open-hearth."

He led me out of the shanty, past three stoves, up an iron staircase, past a blast-furnace, and through a "cast-house." That is not as interesting as I hoped. It is merely a place of many ditches, or run-ways, that lead the molten iron from the furnace to the ladle. Very little iron is ever "cast," since the blast-furnaces here make iron only for the sake of swiftly transporting it, while still hot, to the Bessemer and open-hearth, for further metamorphosis into steel.

We came at last to more stoves, a set of three for No. 4 blast-furnace. Near the middle one was a little group of seven men, three of them with a bar, which they thrust and withdrew constantly in an open door of the stove. Inside were shelving masses and gobs of glowing cinder.

"You work with these feller," Adolph said; and passed out of sight along the stoves.

I watched carefully for a long time, which was a cardinal rule of practice with me on joining up with a new gang. It was best, I thought, to shut up, and study for a spell the characters of the men, the movements and knacks of the job. I think this reserve helped, for the men were first to make advances, and before the day was out, I had a life-history from most of them.

"Where you work, las' job?" asked a little Italian with a thin blond moustache, after he had finished his turn on the crowbar.

"Open-hearth," I said, "third-helper."

"I work three week open-hearth," he said, "too hot, no good."

"Hot all right," I said; "how's this job?"

"Oh, pretty good, this not'ing," he said; "sometime we go in stove, clean 'em up, hot in there like hell. Some day all right, some day no good."

I had been watching the stove, and caught the simple order of movements. Two or three men, with long lunging thrusts, loosened the glowing cinder inside a fire-box; another pulled it out with a hoe into a steel wheelbarrow; another dumped the load on a growing pile of cinder over the edge of the platform. When one of the men disappeared for a chew, I grabbed the wheelbarrow at hauling-out time, and worked into the job.

In fifteen minutes that fire-box was cleared out, and we moved to the next stove. We skipped that; the door was locked and wedged. I learned later that, if we had opened it, the blast (being "on" in the stove) would in all likelihood have killed us. It blows out with sufficient pressure to carry a man forty yards. But the next stove we tackled. I tried the thrusting of the bar this time. The trick is to aim well at a likely crack, thrust in hard and together, and with all the weight, on the bar, spring it up and down till the cinder gives. It was good exercise without strain, and so cool in comparison with open-hearth work that I took real joy in the hot cinder. The heat was comparable to a wood fire, and only occasionally was it necessary to hug close.

We did five stoves, taking the wheelbarrow with us, and carrying it up the steps, when we passed from one level to another. After the five came a lull. Two of the men rolled cigarettes, the rest reinforced a chew that already looked as big as an apple in the cheek. For both these comforting acts

"Honest Scrap" was used, a tobacco that is stringy and dark, and is carried in great bulk, in a paper package.

The men sat on steps or leaned against girders. A short Italian near me, with quick movements, and full of unending talk, looked up and asked the familiar question, "What job you work at last time?"

"Open-hearth," I said.

"How much pay?"

"Forty-five cents an hour."

"No like job?"

"No, like this job better," I returned.

He paused. Then, "What job you work at before open-hearth?"

"Oh," I said, "I was in the army."

His face became alert at once, and interested. The others stopped talking, also, and looked over at me.

"Me have broder in de American army; no in army, mysel'; me one time Italian army. How long time you?"

"Nearly two years," I said.

"Oversea?"

"Yes, but did n't get to front, before war over. No fight," I answered, adopting the abbreviated style, as I sometimes did. It seemed unnecessary and a little discourteous to use a rounded phrase, with all the adorning English particles.

He jumped down from the steps and took up a broom, executing a shoulder arms or two, and the flathand Italian salute, performed with a tremendous air.

"Here," I said, "bayonet."

I took the broomstick, and did the bayonet exercises. The gang stood up and watched with delight, making comments in several languages. Especially the eyes of the Italians danced. The incident left a genial social atmosphere.

Adolph came in from behind one of the stoves as I was concluding a "long point."

"Come on," he said, looking at me with a grin; and when I had followed him, "I show you furnace, li'l bit."

He took me, to a stair-ladder near the skip that ascended to the top of Number 5. For every furnace, a skip carries up the ore and other ingredients for melting inside. It is a funicular-like thing, a continuous belt, with boxes attached, running from the "hopper" at the top of the furnace to the "stockroom" underground.

We started to climb the steps at the left of the belt. There was a little rail between us and the moving boxes of ore.

"See dat," said Adolph, pointing through at the boxes. "Keep head inside," he said, "keep hand inside, cut 'em off quick." He illustrated the amputation, with great vivacity, on his throat and wrists.

It was a climb of five minutes to the furnace-top. We paused to look at the mounting boxes.

"Ore?" I asked.

He nodded.

Pretty soon the iron ceased coming, and a white stone took its place in the boxes.

"What's that?"

"Limestone," he said. "Next come coke. Look."

We were near enough to the top to see the boxes tilt, and the hopper open and swallow the dumping of stone. In a minute or two, we stepped out on the platform on top of the furnace.

Adolph looked at me and grinned. "You smell dat gas?" he asked.

I nodded. He referred to the carbon monoxide that I knew issued from the top of all blast-furnaces.

"You stay li'l bit, pretty soon you drunk," he said.

"Let's not," I returned.

"You stay li'l bit more," he continued, his grin broadening, "pretty soon you dead."

I learned in later days that this was perfectly accurate.

We stood on a little round platform fifteen or twenty feet

BLAST-FURNACE APPRENTICESHIP

across, with the hopper in the centre gobbling iron ore and limestone. A layer of ore dust, an inch thick, covered the flooring, and a faint odor of gas was in the air. Each of the other five furnaces had a similar lookout, and a narrow passageway connected them with the tops of the stoves. The top of these gigantic shafts likewise had a diameter of some fifteen feet; there were little railings about them, and in the centre a trapdoor.

"What's that for?" I asked.

"Go inside to clean 'em out," he returned.

The blast furnace is on the right while the three cylindrical stoves are to the left. About halfway up the incline on the furnace is one of the two "skip" boxes. The stock house is behind the concrete wall. In the foreground is the ore field. Homestead (PA) Works, Carrie Furnace, Rankin, PA, United States Steel Corporation. Photo: Archives of Industrial Society, Wm. J. Gaughan Collection, University of Pittsburgh.

I wondered, with a few flights of imagination, what that job would be like, and remembered that the Italian with the blond moustache had spoken of the duty in uncomplimentary terms.

We could look forth from this eminence and see the whole mill yard, which was nearly a mile in extent. Over the "gas house" a large building I had n't noticed before, the source of gas for the open-hearth — and far to the left, were the Bessemers, spouting red gold against a very blue sky. On their right rose the familiar stacks of the open-hearth. I looked intently at them and wondered what Number 7 did at that moment — front-wall, back-wall, or tapping its periodic deluge of hot steel?

In the foreground, a variety of gables, and then the irregular roof, far beyond, that I knew must be the blooming-mill, because of the interesting yard with the muscular cranes, tossing about bars and shapes and sheets of steel. An immense system of railways everywhere, running down as far as the river bank, where were piles of cinder, and a trainload of ladles moving there to dump. A half-mile away another ironclad cluster of buildings, the tube mill, the nail mill, and the rest, with convenient rails running up to them.

I turned around. Near by, slightly beyond the foot of the skips, was that impressive hill of red dust, the ore pile. Iron ore was being taken away for the skips with one of those spider-like mechanisms that combine crane, derrick, and steam-shovel. It was built hugely, two uprights forty or fifty feet high, at a distance I estimate, of a hundred yards, with their bases secured to railway cars. A crossbeam joined them, which was itself a monorail, along which a man-carrying car ran. From that car dropped chains, attaching themselves at the bottom to the familiar automatic shovel or scoop.

First the whole arrangement moved — the uprights, the crosspiece, and the monorail car — very slowly over the whole hill of ore, to a good spot for digging. Then the monorail car sped to the chosen position, and the shovel fell rapidly into the ore. With a mouthful secure, the chains lifted a little, enough to clear the remaining ore, and the car ran its mouthful to the hill's edge, to dump into special gondolas on railroad tracks. The whole gigantic ore-hill was within easy reach of a single instrument.

"Ought to last a while," I said.

"Will be gone in a month," he returned.

We went down the ladder-steps, and stopped near one of the furnaces. I rather hoped the stove-gang boss would talk. He did.

"Ever work blast-furnace before?" he began.

"No," I said; "I have worked on the open-hearth furnaces a little. But before that I spent about two years in the army."

"Me Austrian army," he said musingly, "fifteen year ago. Sergeant artillery."

I thought about that, and it occurred to me that he retained something of the artillery sergeant still, necessarily adapted a little to the exigencies of American blast-stoves. I found he knew about ordnance, and boasted of Budapest cannon-makers.

"How do you like this country?" I asked.

"America, all right," he said.

"Good country?" I pushed him a little.

"Mak' money America," he explained; "no good live. Old country fine place live."

We developed that a little. We discussed cities. He asked me about London and Paris, and other European cities. Which did I like best, cities over there or American cities? I said American cities. He asked what was the difference. I thought a minute, comparing New York and London. European cities did not have the impressive forty-story edifices of American, and looked puny with four or five.

"Ah," he said, "tall buildings no look good. Budapest good city, no can build over five story."

Here was unlooked-for discrimination. I began feeling provincial. He went on to describe the cleanliness of Budapest, and to contrast it with Pennsylvania cities of his acquaintance. He certainly had me hands down.

He continued: "No can build stack that t'row smoke into neighbor's house. Look at dis place," he said, pointing to Bouton, "look at Pittsburgh."

I said no more, but nodded swift agreement.

He was a little more encouraging about the United States when it came to government.

"You have a man president; that no good, after four year you kick him out. My country sometime get king, that's all right, sometime get damn bad one. No can kick him out."

But he relapsed into censure again when he came to American women. "Women," he said, "in my country do more work than men this country."

"They have more time here," I said, "and don't have to work so hard."

"American women, when you meet 'em, always ask: 'How much money in de pock?' What they do? Dress up, — hat, dress, shoe, — walk all time Main Street. Bah!"

It was a refreshing shock to receive this outspoken critique of America from a Hunky, a Hungarian stove-gang boss of a blast-furnace. I was amused very much by it, except the phrase "America all right mak' money, old country place live." I coupled it up with some talks I had had with men on the open-hearth. America, steel-America, which was all they knew, was very largely a place of long hours, gas, heat, Sunday work, dirty homes, big pay. There was a connection in that, I thought, with the gigantic turnover figures of laborers in steel, the restless moving from job to job that had been growing in recent years so fast. Too many men were treating America as a good place for taking a fortune out of. The impulse toward learning English, building a home, and becoming American, certainly was n't strong in steel-America. But I left these questions in the back of my head, and returned to the stove-gang at Adolph's command.

In a few days I was well in the midst of my gang-novitiate. We got formally introduced by name one day in front of No. 12 stove. The little Italian with the black moustache said: "What's your name?"

"Charlie," I said, knowing that first names were the thing.

"All right," he said, "that's Jimmy, Tony, Joe. Mike not here. You know Mike? Slavish. John, that's me. That's John too wid de bar. — Hey!" with an arresting yell, that made the others look up, *"Dis is Charlie!"*

I became a part of an exclusive group of seven men, who had worked together for about two years. There is a cohesiveness and a structure of tradition about a semipermanent mill-group of this sort that marks it off from the casual-labor gang. The physical surroundings remain unaltered, and methods and ways of thought grow up upon them. I was struck by the amount of character a man laid bare in twelve hours of common labor. There are habits of temper, of cunning and strength, of generosity and comradeship, of indifference, that it is capable of throwing into relief beyond any a prior reasoning. It begins by being extensively intimate in personal and physical ways; you know every man's idiosyncrasies in handling a sledge or a bar or a shovel, and the expression of his face under all phases of a week's work; you know naturally the various garments he wears on all parts of his body. You proceed to acquaint yourself, as the work throws up opportunity, with the mannerisms and qualities of his spirit. It is astonishing, with the barrier of a different language, only partly broken down by a dialect-

American, how little is ultimately concealed or kept out of the common understanding.

I was impressed by the precise practices established in doing the work. Every motion and every interval of the job had been selected by long trial. If you did n't think the formula best, try it out. Many considerations went into its selection — to-day's fatigue, to-morrow's and next month's. It had an eye for gas effect, for the boss's peculiar character, and for all material obstacles, many of which were far from obvious.

When the flue dust had been removed from the blast-stoves, I found wheeling and dumping it an easy and congenial set of movements, and consequently took off my loads at a great speed.

At once I became a target. "Tak' it eas' — What's the matter with you; tak' it eas'."

John — Slovene, and Stoic — put in an explanation: "Me work on this job two year, me know; take it easy. You have plenty work to do."

"Take it easy," I said, "and no get tired, eh? feel good every day?"

"You no can feel good every day," he amended quickly. "Gas bad, make your stomach bad."

So I slowed up on my wheelbarrow loads, sat on the handles, and spat and talked, till I found I was going too slow. There was a work-rhythm that was neither a dawdle nor a drive; if you expected any comfort in your gang life of twelve hours daily, you had best discover and obey its laws. It might be, from several points of view, an incorrect rhythm, but, at all events, it was a part of the gang *mores*. And some of its inward reasonableness often appeared before the day was out, or the month, or the year.

Everybody wore good clothes to work, and changed in the shanty to their furnace outfit. I usually came in a brown suit, which had been out in the rain a good many times and was fairly shapeless. One day I entered the mill in a gray suit, which fitted and was moderately pressed.

At the dinner-bucket hour in the shanty, I was asked by John the Italian: "How much you pay for suit, Charlie?"

I was embarrassed, fearing vaguely explanations that might have to follow a declaration of price. I suddenly recalled the fact that the suit had been given me by my brother, so that I did n't know the price, and said so.

"My brother give me suit, I don't know how much he pay," I said. That dumped me into another quandary.

"What job your brother have?" I was immediately asked.

I thought a moment and answered truthfully again.

"My brother, priest," I said.

That arrested immediate attention, and I was looked at with respect and curiosity.

Tony finally said, "Why you no be priest, Charlie?"

"Oh," I answered, laughing, "I run away; I like raise hell too much be priest." This was pretty accurate, too.

"O Charlie!" they bellowed.

After that, the gang were friends to the death.

VII

DUST, HEAT, AND COMRADESHIP

One day I was promoted to stove-tender or hot-blast man on Number 6.

The keeper of the furnace was a negro. When he was rebuilding the runways for the tapped metal, I noticed that his movements were sure and practised. He patted and shaped the mud-clay in the runway, like a potter moulding a vessel. When it was tap time, he bored the tap hole with the electric drill easily and neatly; when the metal flowed, he knew the exact moment to lift the gates for drawing away slag. I watched him to see how he managed the four white men that worked for him. They were Austrians, and I found they joked together and showed no resentment of status. Commands were given with a nod or gesture. With the Americans on the furnace, the relation was the traditional one. The negro was light and seemed too slightly built for the job, but he performed it very efficiently, and so did his gang.

The blower was Old McLanahan, a man somewhere between thirty-five and sixty. A long, successful life of inebriety had given him a certain resignation to the ills of man, and enabled him to keep the heart of a *viveur* throughout his life. His skin appeared thrown like a bag over an assemblage of loosely fitted bones — the only considerable part of him being a paunch which coursed forward into a moderate point.

He was rather proud of being a blower on furnace No. 6. After the slag had been sampled he said: "Where d'ye eat, boy?"

"I eat at Mrs. Farrell's."

"How much?"

"Seven a week."

"Too much. Pretty goddam good is it?"

"Damn good food," I said.

"Is Mrs. Farrell a widder woman?"

"No," I said, "she's not."

"Well," he said, if you hear of a damn fine little widder woman, let me know will yer?"

An employee working on the "monkey" or slag runner. Slag floats on the molten iron, so a separate tap hole for the slag was provided. This was called the "monkey" or cinder notch. Homestead (PA) Works, Carrie Furnace, Rankin, PA, Mar. 5, 1954, United States Steel Corporation. Photo: Archives of Industrial Society, Wm. J. Gaughan Collection, University of Pittsburgh.

DUST, HEAT, AND COMRADESHIP

"Sure," I said.

"I'm lookin' for a place ter board, and most of all I'm lookin' for a little widder woman ter honor wid holy matrimony."

After tapping that morning at 8.00, McLanahan took a silver dollar out of his pocket. "If it comes heads," he said, "I'm goin' out to-night, see, I'm goin out ter find a woman."

He flipped the coin and it fell tails. "Don't count," said he, "two out of three."

This flip fell heads.

"Hah," he said, "if this comes heads, I'm goin' out to-night ter find a woman."

It fell tails.

"Hell!" he said, "Don't count, flipped it with the wrong hand."

He kept this up all day. Finally at 5.30 the coin came heads. He picked the coin up and put it in his pocket.

"Goin' out, to-night," he said.

"Boss wants to keep Number 6 lookin' right down below, and clean out all that flue dust."

I shoveled between the stone arches of the furnace base, that curved overhead like the culverts of a bridge. Sometimes the flue dust was wet and clotted with mud, and came up in cakes on the shovel; sometimes it was light, and flew in your nose and eyes. I made a pile of it six feet high, and shaped it into a brick-red pyramid with my shovel. I washed the arches white with a hose.

"Change 'em before we tap," McLanahan ordered, nodding at the stoves.

I went among the rangy hundred-foot shafts with a certain sense of control over great forces. Every set differs in its special crankiness. Number 9's have stiff-working valves, but are powerful heaters; Number 8's are cool stoves, but their valves slide genially into place. I always a little dreaded "blowing her off." Resting my arms on the edge of the wheel, and grabbing the top with my hands, I wrenched it over to the left, and the blast began. The immense volumes of compressed air escaped with a gradually accelerated blare. I gritted my teeth a little, and my ears sang.

Then came "putting on the gas." I climbed to a little platform near the combustion chamber, and with a hunk of iron scrap for hammer, knocked out some wedges that held tight a door. By now I knew just the pressure for making the iron slab creep on its rollers. I braced my feet and pulled with back and arms.

Through the door, the combustion chamber glowed red. I went down the steps and slowly turned the gas-pipe crank, bringing an eight-inch pipe close to the red opening. I dodged the back flare as it ignited.

When the "new" stove was on, and the "old" one lit for reheating, I went to the pyrometer shanty. In a little hut among the furnaces were tell-tale discs, that let you know if you were keeping your heat right. I found my heat curve was smooth with only a tiny lump. . . . Two Hunkies were inside the shanty.

"Nine-thirty," said one.

"How do you know?" I asked.

He pointed to the end of the curve on the disc, that was opposite the 9.30 mark on the circumference.

"Saves me a watch," he said, with a grin.

After supper that evening, I mended a sleeve of my shirt that had been torn on a piece of cinder in the cast-house. Sounds of conversation were rising from the porch. I went out and found Mr. Farrell sitting in a rocker with one leg on the railing and his face screwed into in attitude of thinking. Mrs. Farrell, having done the dishes, had come out to knit, and a lanky visitor, who leaned uncomfortably against the railing, was doing the talking. The conversation was political.

"Before I came to this town, nobody had the guts to vote Democratic," said the visitor. "I'm from Democratic parts," he went on, "and when I first come here I used to go round. 'Come, come,' I said, 'you fellers is Democrats, you know you is. Sign up.' 'We know it,' they 'd say, 'but we can't afford ter, there 's the wife and kids — we can't afford ter, we've got a job and we're goin' to keep it.' That's how bad it was."

"You mean — "

"I mean you voted with the Company or pretty quick you moved out of Bouton, for you had n't any job to work at.... I used ter work at glass blowin', that's a real business — "

"Mr. Herder is always telling us how much better the glass business is than the steel business," said Mrs. Farrell. "You'll have to get used to that." She gave everybody a smoothing-out smile.

It was fun when you could pick up "dope," in the course of a morning's sweat. I learned one Sunday a few pointers about judging conditions through the peepholes. If there is a lot of movement, your furnace is O.K. If the cinder begins to settle into the tuyere, your furnace is cold. If she looks reddish, cold; blue, O.K. Don't be fooled by different colored glasses in the peepholes.

One day we kept the stoves on "all heat" for the furnace was cold. "All you can give her, goddam it," McLanahan said, looking through the peepholes. McLanahan was always a little ridiculous. Anxiety made him hop about and waddle from peephole to peephole, like a hen looking for grain.

I heaved on the hot-blast chain, and the indicator climbed.

We had a pleasant, light brown chocolaty slag that day, which meant good iron. When the metal runs out with large white speckles, she has too much sulphur; when she smokes, you 'll get good iron.

The other day they had too large a load of ore for the coke and stone in her.

"Sledge!" yelled the keeper.

A cinder-snapper brought up two, and held the bar while the keeper and first-helper sledged. They worked well, and I watched with fascination the hammer head whirl dizzily, and land true at the bar.

At last the liquid slag broke through, jet-black as if it were molten coal, flowing thickly down the clay spout. The clay notch was hammered and eaten away, and had to be remade.

I watched the stove-tender on Number 7 as he opened the cold-air valve. His motions were exactly calculated — the precise blow, to an ounce, to loosen that wedge.

Drilling the tap hole. Ken Kobus photo.

DUST, HEAT, AND COMRADESHIP

"How long have you been stove-tender?" I asked.

"Ten years," he said.

"Go down to the stockroom and tell the skip-man, one more coke," said McLanahan.

I was glad to get a glimpse of that part of the blast-furnace operation. Gondola cars bring up ore and the other ingredients of blast-furnace digestion, and run over tracks with gaps between the sleepers. The cars, by means of their collapsible bottoms, drop the loads down through, and the material falls into an underground "stockroom."

I entered it by climbing down two ladders, and found the skip-man at the base of one of the endless chains. The chamber had the appearance of a mine gallery de luxe. I looked at the tons of ore moving upward neatly, efficiently. What an

The cast house at Aliquippa. Iron and slag flow through their respective "runners." Aliquippa Works, Jones and Laughlin Steel Co. Photo: Don Inman Collection, Beaver County Industrial Museum, Geneva College, Beaver Falls, PA.

incalculable saving of labor and time, this endless chain affair with its continually moving boxes, over the old manner of hoisting painfully, in few-pound lots, by hand!

I gave McLanahan's order to the skip-man and went up the ladders.

You 've got to tap, "when the iron's right," and when a little later the keeper held the steam drill in front of the mud wall of the tap hole, the steam stayed at home. There was no time for a steam-fitter.

Young Lonergan and I beat it for the electric drill. It was heavy enough to make us waddle as we carried it on the run.

"That's bludy funny," said McLanahan. The electric drill would n't electrify. A hurry call followed for the electrician. He smiled benignly while twelve sweaty men looked on. And in thirty seconds he fixed the connection, and we tapped in time to save the iron.

Iron being cast into a torpedo style ladle. Aliquippa Works, Jones and Laughlin Steel Company. Photo: Don Inman Collection, Beaver County Industrial Museum, Geneva College, Beaver Falls, PA.

DUST, HEAT, AND COMRADESHIP

When the drill had almost bored through the hard mud in the tap hole, the keeper shoved in a crowbar, and a couple of helpers sledged rhythmically for one minute. Then the molten iron broke the mud into bits, and tumbled out. Little sheets of flame from the slag skated along the top of the red river. It rose in the runway with bubbles and smoke on top. The keeper grabbed a scraper — an exaggerated hoe — and started the slag through a side ditch.

"Now try it," said Old Mac.

By then, I had the test spoon ready, scooped up a bubbling ten pounds, carried it carefully, and poured it into two moulds.

When I had broken the little ingots, still red, Mac said, "Too much sulphur."

By now the metal stream had run to the edge of the cast-house and was falling spatteringly into a ladle ten feet below.

In this view the slag is running into "cinder pots," sometimes called "thimbles." Aliquippa Works, Jones and Laughlin Steel Company. Photo: Don Inman Collection, Beaver County Industrial Museum, Geneva College, Beaver Falls, PA.

Somebody said, "Whoop!" The negro keeper opened the iron gate of a new runway, and the metal rolled on its way to a second ladle. There were five to fill, each on a railway car. I noticed the switch engine was getting ready to drag the trainload of molten metal to the Bessemer.

"Heow!" out of Old Lonergan's throat. The bottom of one ladle had fallen out and was letting down molten iron on the

Five furnaces in cast simultaneously. Aliquippa Works, Jones and Laughlin Steel Company. Photo: Don Inman Collection, Beaver County Industrial Museum, Geneva College, Beaver Falls, PA.

track. There was nothing to do but watch it. We did that. It covered the track like a red blood-clot, and ran off sizzling, and curdling in the sand. It cooled, blackened, and clotted over one rail — about 10,000 pounds.

"Who clean dat up?" I heard a Sicilian cinder-snapper say with a blank smile.

After the furnaceful of metal had all flowed forth, we prepared to plug that tap.

I went over to the other side of the tap hole, and picked up a piece of sheet iron. A shallow puddle of iron was still molten in the runway. The tap hole was crusting over with cooling iron, still aglow. I dropped the sheet iron over the runway. The helpers came up behind and dropped others.

"Hey, you," said the keeper summoning a helper. They swung out the "mud gun" on a kind of crane, and pointed its muzzle into the glowing aperture. It was a real gun, looked like a six-inch fieldpiece, but fired projectiles of mud by steam instead of powder.

"Quick," said the keeper.

I pushed a wheelbarrow towering with mud up to the sheet iron; then, with a long scoop-shovel standing against the furnace, shoveled mud in the gun. The keeper stood almost over the runway with only the rapidly heating sheet-iron between himself and the liquid-metal puddle beneath. He operated a little lever that shot mud charges by steam into the hole. Every time he shot the gun, I took a new scoop of mud. We worked as fast as our arms let us. Some of the helpers kick at this part of their duties, but it is cooler, by several degrees than the open-hearth, and thinking of those sizzling nights lightens it for me. Besides, it has excitement and requires a streak of skill.

I spent several days with young Lonergan helping the water-tender, Ralph.

"Water connections damn important thing," said Lonergan. I was beginning to see why. The whole wall of the great cone-shaped furnace was covered with cooling water-conduits. Without these the furnace would melt away.

We ranged from furnace to furnace, climbing up to a platform that ran around the fattest part and spending long quarter-hours on our bellies unscrewing valves. There was always something leaking. Ralph could come and take a look at the furnace, and send us after tools.

"Ralph's all right," said Lonergan, "has new names though for everything. Does n't call a goddam wrench a wrench, calls it a 'jigger.' Have to learn all your tools over again by his goddam Hunky names."

Note the mud gun and tap drill on either side of the runner. Aliquippa Works, Jones and Laughlin Steel Company. Photo: Don Inman Collection, Beaver County Industrial Museum, Geneva College, Beaver Falls, PA.

Young Lonergan was very "white" to me, as they say. "I'll show you how to clean that peephole." And he grabs a cleaning rod, and imparts the knack of knocking cinder out of that important little observation post.

"I used to work stove-tender," he explained.

"If you want to know anything ask Dippy, he'll talk, don't McLanahan, he don't know he's livin'.

... Have a chew?"

"No, I'll smoke."

One day we had been discussing the bosses, and how they had got their start, till the talk drifted to young Lonergan and his own very typical career of youth.

"Used to work on the open-hearth," he began. "I used to test the metal — you know in the little shanty where 'Whiskers' is now. Chemist!" he grinned.

"Then, by God, I went to work in the blooming-mill, chasing steel — you know; keepin' track of all the ingots comin' in. A hell of a job — by God you did n't stop a second — you knew you'd be workin', boy, when you pulled out in the mornin'. I worked my head off at that job.

"Then I fought with Towers. He gave me a week. After I came back I had another run-in.... When I carried my bucket out o' that place, I was off work entirely. Did n't go to work for three months, thought I never would work again.

"But after a hell of a spell, gotta job, pipe mill New Naples — eight hours — a good job, but the mill's shut down now. Then the suckers drafted me. Balloon comp'ny a bloody year and a half."

There followed a very vast series of parties in the army, and explicit views on all the officers he'd had.

There was usually a new army story whenever I met him. He was extraordinarily clever in getting away with A. W. O. L.'s.

"When I got my discharge, father wanted me to come to work here, so I did. Worked on those stoves where you are, for a while — stove-tender helper, then stove-tender. Then I got this job. . . . Don't you chew? . . . I'll lose it too if I take many more days off for sickness. Last time I was 'sick'" — he grinned — "Bert Cahill and the bunch and I took three skirts in Bill's car to Monaca. Had six quarts of damn good whiskey. I was out a week. Ralph says, when I come back: 'Pretty damn sick, you!' But to hell with 'em! I'm not afraid of my job."

That little blower called Dippy, I found, knew the furnace game in all its phases with great practical thoroughness. I used to try to get chances of talking with him on questions of technique.

"What about those jobs in the cast-house?" I said one day, "the helper's jobs? Is n't it a good thing to know about those if you're learning the iron game?"

"You don't want to work there," he said quickly, "only Hunkies work on those jobs, they're too damn dirty and too damn hot for a 'white' man."

So I got thinking over the "Hunky" business, and several other conversations came into my mind. Dick Reber, senior melter on the open-hearth, had once said, "There are a few of these Hunkies that are all right, and damn few. If I had my way, I'd ship the whole lot back to where they came from."

Then I thought of the incident of my getting chosen from the pit for floor work on the furnaces. Several times Pete, who was a Russian, discriminated against me in favor of Russians. Until Dick came along and began discriminating in my favor against the Hunkies.

How many Hunkies have risen to foremen's jobs, I thought, in the two departments where I have worked? One in the open-hearth — a fellow who "stuck with the company" in the

Homestead Strike — and none on the blast-furnaces except Adolph, the stove-gang boss.

My recollections were broken into by a call for violent action.

"Cooler," yelled McLanahan, his voice going up into a husky shriek.

That meant molten iron inside, melting the cone-shaped water-chamber around the blast pipe. If let alone, the cooling at that place would cease, and in a short time there would follow an escape of molten metal.

"Cooler!" yelled on a blast-furnace means "Hurry like hell."

I grabbed a wrench to take the nut off the "bridle" — the first step in taking out a sort of outside cooler, the tuyere.

"Bar," said the Serbian stove-tender very quietly, picking up a specially curved one, and McLanahan took the other end.

Somebody knocked out some keys with a sledge, and the blowpipe fell on the curved bar, making the holders of it grunt. They took it off fast, for the instant the thing loosens, a flame shoots through the hole and licks its edges.

Then the tuyere comes loose with a few strokes of a pull bar. All of these moves are fast; a tuyere goes bad every other day and men work fast like soldiers at a gun drill.

But coolers don't break a lead but once in three months or so; and the cone's heavier, the gang bigger, there's less efficiency and more holler and sweat.

When the pull bar gets into action it looks a little like a mediæval mob with a battering ram. A "pull-bar" is a tool designed to translate the muscle of many men into pull, on a small gripping edge against which sledging is impossible. At one end a thick hook grips the edge of the cooler, at the other a weight is brought against a flange that runs around the bar. Everybody on the gang has a piece of rope attaching to that weight.

The stove gang moving between stoves Thirteen and Fourteen were caught and brought into this for muscle, and a couple of passing millwrights drafted.

"Hold up the goddam end," from Steve, boss by common consent.

"A little beef this time!" from a blower. "What the hell's the matter, *sick?*"

We all swear between breaths, and take a grip higher on the rope — the weight cracks the flange again, and makes the bar shiver.

When the new cooler, which resembles more nearly a gigantic flower pot, without any bottom, than anything else, is in place, there's a cry of: "Big Dolly!"

That involves four or five men, lifting a kind of ramrod with a square hammer-end, from the rack, and lugging it to the cooler.

I get near the ramming end this time; Tony is near me on the other side. Together we hold the hammer against the cooler. As the end strikes, the jar goes back through the men's hands.

"Now top."

Arms raise the bar painfully, and hold it poised a little unsteadily, sway back, tense and drive.

"Hold it, hold it on the cooler, goddam you."

Tony and I had let our arms shake a fraction, and the hammer fell glancing on the cooler's edge.

"Now!"

Seated this time. Arms relax and stretch.

When things are ready, Adolph makes the water connections.

"Hold de goddam shovel, what you t'ink, I burn up."

A cinder-snapper holds a shovel in front of the hole to keep the flame from his hands.

"All right, all right."

The job's done; the millwrights pick up their tools, and the stove gang moves off leisurely to their cleaning. I hear the superintendent talking with a blower near the sample box.

"They did that in pretty good time," he says.

DUST, HEAT, AND COMRADESHIP

I used to eat my lunch and kept my clothes in a little brick shanty near Number 4, sharing it with the Italians of the stove gang. Although by the bosses' arrangement it was a mixed gang, Italian and Slav, the mixture did not extend to shanty arrangements, and race lines prevailed. I felt that I should learn low Italian in a few weeks if I continued with this group; the flow of it against my ear drums was incessant and some of it

At this level of the furnace we see the hot blast piping. The elbow-shaped pipes, marked 18 and 1, are called penstocks. The small circles below the numbers are the peepholes for the furnace. The horizontal pipe on the lower end of the penstock is the blowpipe. At the end of the blowpipe is the water cooled tuyere, which cannot be seen in the photo. The circular opening around the blowpipe is the cooler. Each of these parts weighs hundreds of pounds and must be changed quickly while the furnace is still hot. Aliquippa Works, Jones and Laughlin Steel Company. Photo: Don Inman Collection, Beaver County Industrial Museum, Geneva College, Beaver Falls, PA.

had already forced an entrance. Besides I was learning a great deal about: how to live, what to wear on your head, on your feet, and next your skin; where to get it — good material to resist the blast-furnace, and cheap as well; wisdom in eating and drinking, and saving money, in resting, in working, in getting a job and keeping it.

There was a whole store of industrial *mores*. In some respects the ways of living of these workmen seemed as rooted and traditional as the manners of monarchs, and as wise. I won considerable merit, when I brought in a kersey cap that I got for seventy-five cents, and lost much when I reluctantly admitted the price of my brown suit.

Everyone on the gang performed the washing up after work with the greatest thoroughness and success. They devoted minute attention to the appearance of clothes worn home. Rips and holes got a neat patch at once, and shoes were tapped at the proper period —before holes appeared. I have seen only one or two men in the mill who were not clean in their going-home clothes.

I talked to John one day on the subject of neatness. He asked, "You have to clean up good in the army?"

I dilated on the necessity of policing when wearing khaki.

He said: "Man that no look neat, no good. I no like him, girls no look at him. Bah!"

I was almost always offered some food from the bursting dinner buckets of my friends: a tomato, some sausage, a green pepper, some lettuce and cucumbers. I accepted gladly for it was always superior to my restaurant provender.

Tony told me one day that Jimmy had come over "too late from old country, to learn speak English and be American." He was thirty-one years old. He was going back this Christmas. And Tony was going too, but just for a visit. They were going to Rome. We had talked it over a good many times, all Italy in fact, people, women, farms. Tony turned to me: "You come Italy with Jimmy and me this Christmas? We go see Rome."

I assented quickly, wishing I somehow could, and was extraordinarily proud of that invitation.

I must not forget the occasion of the green pepper. One noon I sat beside Jimmy during the lunch hour. The whole Italian wing were together, sitting on benches in the brick shanty. Jimmy reached among the loaves of bread in his bucket, and hauled out a green pepper as big as an orange. He offered it to me and I accepted.

Treating it like my old friends the stuffed peppers, I bit deep. The whole shanty watched eagerly for results. I hadn't reckoned its raw strength and instantly felt like a blast-furnace on all heat. Despite all efforts I couldn't keep my face in shape, or resist putting out the fire with the water jug. The pleasure I furnished the Roman mob was enormous.

After that I learned to eat green peppers rationally and agree with my friends that they are beneficial. Beyond their health qualities they have an economic justification. With their help you can make a meal of cheap dry bread. Plain and unbuttered it costs you but six cents a half loaf which is a full meal, and hot green peppers will compel you to stow it away in self-defense. As Tony phrases it:—

"Pepper, make you eat bread like hell!"

Tony thinks that Americans eat too much that is sweet; it makes them logy and sleepy. I think he is right. Joe claims that the people in America do not know how to make bread; the wheat he says is cut when it is too green. The gang, of course, bring Italian bread in their buckets. It is certain that the American lunch of a soggy sandwich and piece of pie leaves a man heavy for the afternoon. The average dinner bucket in the shanty contains: a loaf of bread, a piece of meat, — lamb, beef, chicken, or sausage, — three or four green peppers, a couple of tomatoes, a bunch of grapes, and some vegetable mixture like tomatoes chopped with cucumbers and lettuce.

One day the gang got absorbed in stunts, climbing a ladder with the hands, giving a complete twist to a hammer with grip the same, the usual turning trick of a broomstick held to the floor, etc. My contribution was squatting slowly on the right leg with the left stiff and parallel with the floor. John complained of a lame thigh for three days after, I am gratified to say.

With Tony I occasionally picked a wrestling quarrel; he has a terrific grip and one day very nearly squeezed the life out of me in a fit of playfulness. I called him "Orso" afterward for his squeezing attribute. Tony's make-up includes a sense of humor. One day when he had rolled about on the floor in front of Number 3, he said: "Ain't you 'shamed, Charlie, you young man, fight old man like me. You twenty-two, twenty-three, me thirty-seven!"

Tony could put me beyond this vale of tears with his left hand.

VIII

I TAKE A DAY OFF

I DECIDED on a day off. John had lately taken one for the festival at New Naples, and had come in to work the next morning with the wine still at festivals in his head. Sitting atop the blast-furnaces the other day, looking at the blue rivers and the three hills, and speculating about men going down to the sea in ships — because of the fat river-boat we could see — had made me sicken of the smell of flue-dust. I decided to take a day off.

Sometimes the foreman, when you got back after cutting a turn, would say, "I don't believe you want this job; you like loafing better; I'll give it to Jimmy." But with a seven-day week, only the mean ones hollered. Men took an occasional holiday.

I ate breakfast with a very conscious leisure at George's, putting down scrambled eggs, at 8.00 o'clock, instead of the coffee and toast at 5.15 A.M.

"No work to-day," said George; "lotza mon', eh?"

"Wrong," said I.

"Mebbe you see best girl to-day."

"Guess again."

"Married?"

"No."

"Mr. Vincent's wife is sick," said George, changing the subject.

"Oh, I'm sorry."

"He no work to-day; come in here for breakfast ten minutes before you."

Vincent was a young American, twenty-one or two, whose brother I had known in college. He had not gone himself, but took a straw boss's job in the pipe mill. He had married six months before, and his wife lived with him in two rooms in Bickford Lodge

— the other hotel in Bouton. We went to the movies together sometimes, and often met for supper at the Greek's.

I looked for Vincent, and found him reading the "Saturday Evening Post" in the front room.

"Elizabeth is sick," he explained. "I'm sticking around to-day."

We fell to talking mill.

"What hours do you work now?" I asked.

"Six to six."

"You get up at five."

"Yes, about that."

"That's not true, Philip," came over the transom from the sick room. "I set the alarm at four-thirty, Phil sleeps till five-thirty, drinks one cup of coffee, leaves his eggs, and catches the twenty-of-six car."

"You now have the story," said Phil. "It's a stinking long day, is n't it?"

"Phil has it all figured out," Elizabeth shouted from the back room. "From six to nine, he pays his rent — "

"Yes, I've figured it that way," he said. "The money I earn between nine and one is enough to pay my day's board and my wife's; one to three is clothes and shoes; three to five, all other expenses; five to six I work for myself!"

"That's bully; I think I'll figure mine."

"But there aren't any evenings, are there," he went on, "or any Sundays?"

Suddenly he looked up at the chandelier. "See all the pipes in that," he said; "I find pipes and tubes everywhere, since I've worked in the mill. It's darn interesting to pick them out. The radiator in this room is made of pipe, see; the bed in the back room; notice those banisters outside. I see them everywhere I look. If I had a little money, I'd put it in a pipe mill. 'S money in that game, once you get the market; Coglin and I have it all doped out."

I TAKE A DAY OFF

For fifteen minutes, Phil's enthusiasm for pipe-manufacture built the mills of the future.

Toward noon I went to George's. The pit crane-man, Herb, was there, eating George's roast beef and boiled potato, and looking half asleep.

"I'll fire you," I said.

"I'm on nights this week," he returned with a slow smile; "I could n't sleep, so I thought I'd get up and eat some. Besides, I've got to go to the bank. You 're with the blast-furnaces now, huh?"

"Yes."

"Like 'em?"

"Yes, I think I 'll like blast-furnace work," I said, "if I get to be stove-tender or something. Good boss, Beck."

"They say so. Pete 's as crabby as ever in our place. He fired one of the second-helpers last week, Eric — d'you know him? Used to come in drunk every day, worked for Jock on Eight."

"That 's too bad," I said; "he gave everyone a good time. Let me tell you how I amuse the gang on the blast-furnace. You know the way they break ingots for a test on the open-hearth?"

"Yes."

"It's not like that with us. I gave everybody on Five a treat because I thought it was."

Herb looked interested.

"Of course, on the open-hearth you pick them up with a tongs, when they're red-hot, and cool them in water."

Herb nodded.

"So there are always halves of test-ingots on the floor, *cold*. On the blast-furnace the stove-tender pours the test and knocks it out of the mould. Iron breaks easier than steel, so he never bothers to cool the ingot, but breaks it red-hot. Last Wednesday I wander up from the stoves when the furnace is ready to tap. The blower kicks busted halves of a test-ingot out of the way, and somebody says, 'A little too much sulphur.' I'm ambitious to learn iron smelting, too, and think I'll study the fracture. I walk in front of the blower and pick up the test.

Herb grinned.

"It was n't red-hot," I went on; "but it had blackened over — *just*. I dropped it, and snapped my hand three feet behind me. The blower, the stove-tender, the first, second, and third helpers, and the assistant superintendent, who were all gathered, enjoyed the thing all over the place for several minutes. It gave them a good time for the afternoon."

When I left Herb, I took a walk through the Greek and Slavic quarters, and stopped a while on Superintendent's Hill, to study the graded superiority of foremen and superintendents. There were excellent little houses here, though too young and new to express any other character than moderate prosperity. Perhaps it was an ungracious thing to demand more.

I walked on, past farms, and up and down considerable hills. I lay down on the ground, in high grass, under apple trees which were near a tumble-down stone wall. It was enormously satisfactory to lie in the high grass, under an apple tree, listening to the small August noises — for a swift hour and a half.

After supper, I wanted badly to take a look at furnace fires against a night sky, and stepped out alone to do it. Close to the railroad station I set foot on the hill, and climbed past a Greek hotel and staggering tenements to a ridge. From there I could look over multitudinous roofs to the flat spaces by the river, where the mills roared and shone.

I heard heavy things dropped here and there over acres of plate flooring; they melted into a roar. The even whirr of the power house increased it, and the shrieks of machinery gave it a streaky quality. There were staccato punctuations, of course, by the whistles, and when a distant "blaw" came to me, I

I TAKE A DAY OFF

thought how loudly it drove into the ears of the hot-blast man turning his wheel by a stove. But it was mostly the summed-up roar that occupied your head — an insistent thing, that made you excited and weary at the same time. The mills had been running for ten years; they always had a night-shift in Bouton.

It is easy to get excited about a steel-mill sky at night. I like to look at them. There were n't many lights at the nail mill but just enough to show broken outlines of a sheet-iron village

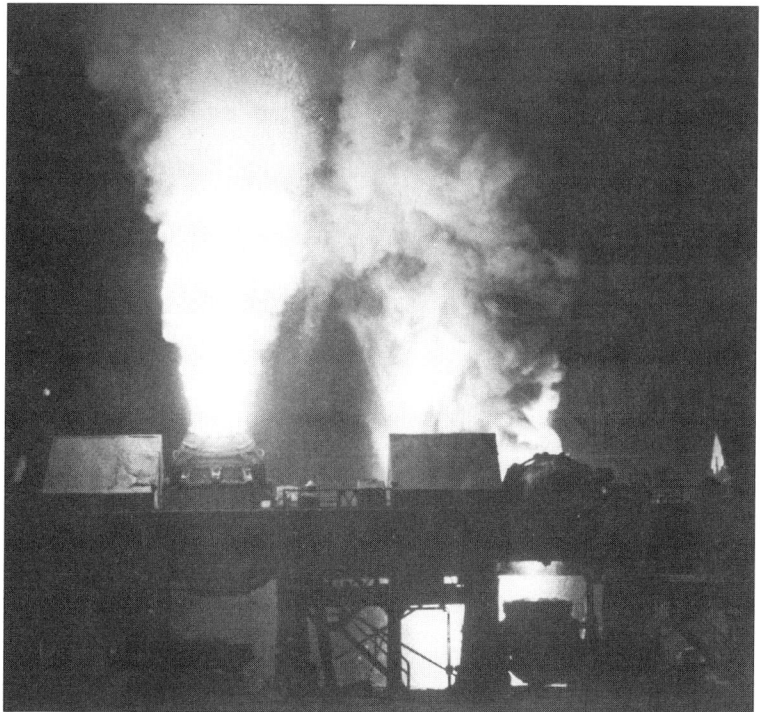

"I could see the Bessemer converter pouring a fluid rope of white light; I knew it for a stream the thickness of a hydrant. A rusty, glowing cloud rose over the converter, changing always, and turning that patch of sky into gold." Provenance unknown. Ken Kobus collection.

there. The rolling-mills gave some of the brightness of hot billets through the windows, and over the stacks of the open-hearth were sparks. By closing my eyes, I could see curdling flame in the belly of Number 7. The open-hearth fires showed themselves, a confused glow under a tin roof.

Some little light came on the mills out of the night itself, though thin clouds kept washing the face of the moon, and now and then a blast-furnace got into the moonlight and looked perfectly confused with its pipe labyrinth and its stoves.

From where I stood, I could see the Bessemer converter pouring a fluid rope of white light; I knew it for a stream the thickness of a hydrant. A rusty, glowing cloud rose over the converter, changing always, and turning that patch of sky into gold. The pattern of smoke the blower knows like a textbook, and follows the progress of his steel by the color of the cloud.

My mind swept over many memories as I looked at the yellow fire of the Bessemers. There was no order or arrangement in them. They were a stream, thick in some passages, shallow in others, with scraps of all sorts riding over the top. One scrap was the price the Wop cobbler charged for soling, and another, Dick's words when he damned me for forgetting a bag of coal. Then there were things that wrung me and made the palms of my hands wet, as if thoughts went over nerves and not brain.

I looked over at the eight stacks of the open-hearth, closed my eyes, and saw Seven tapping. The second-helper broke the mud stoppage with his "picker," and liquid steel belched. Pete held up two fingers. Stanley the Pole was third-helper with me. We shoveled in the two piles. I could feel heat in my nose and throat and sparks light on the blue handkerchief I had tied around my neck. We cooled off in a breeze between the two furnaces, and as we caught our breath, watched Herb swing the ladleful, over the moulds for pouring.

I TAKE A DAY OFF 131

I lived through the dragged hours in the morning of a long turn. Between two and four is worst — I remembered "fixing the spout" with Nick at three — wheelbarrow loads of mud and dolomite — a pitched battle with sleep —

At intervals in my memories, I grew conscious of the steady roar the mills sent me from the river; then forgot it, quite.

Finished ladles of iron came into mind, and I tried to follow in the dark the path they would take along tracks to the Bessemer. Thick red ingots of steel, big as gravestones, I knew, were coming from "soaking pits" to rolls, and getting flattened into blooms and billets. I could see trainloads of even steel moving out of the freight yard to become the steel framework of the world.

"It is perfectly certain that civilization is kept from slipping, by a battle," I said to myself, beginning a line of thought.

"Thick red ingots of steel big as grave stones." Aliquippa Works, Jones and Laughlin Steel Company. Photo: Don Inman Collection, Beaver County Industrial Museum, Geneva College, Beaver Falls, PA.

An express train shot into view in the black valley at my feet, and passed the Bouton station, with that quickly accelerating screech that motion gives. I thought of the steel in the locomotive and thought it back quickly into sheets, bars, blooms, back then into the monumental ingots as they stood, fiery, from the open-hearth pouring, against a night sky. Then the glow left, and went out of my thinking. Each ingot became a number of wheelbarrow loads of mud, pushed over a rough floor, Fred's judgment of the carbon content, and his watching through furnace peepholes. The ladlefuls ceased as steel, becoming thirty minutes' sledging through stoppage for four men, the weight of manganese in my shovel, and the clatter of

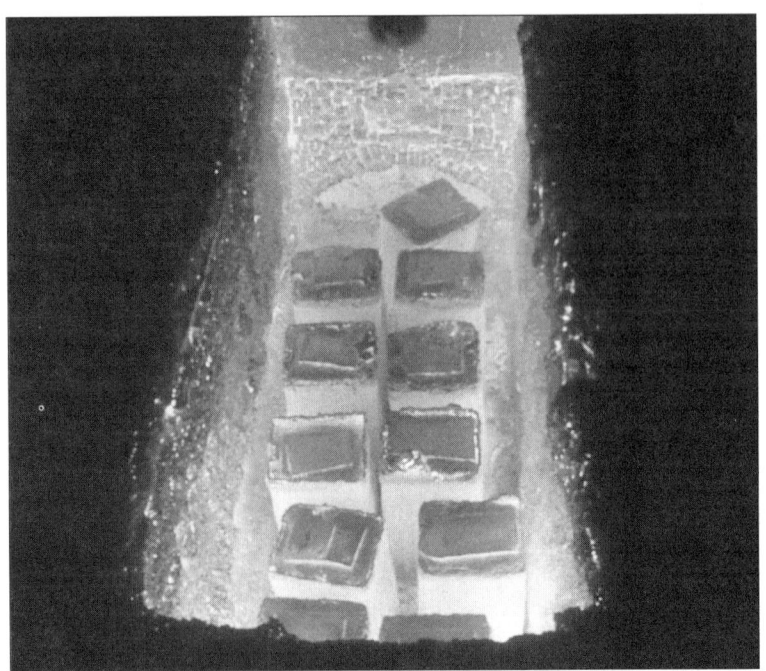

Ingots in a soaking pit. Aliquippa Works, Jones and Laughlin Steel Company. Photo: Don Inman Collection, Beaver County Industrial Museum, Geneva College, Beaver Falls, PA.

the pieces that hit the rail, sparks on my neck burning through a blue handkerchief and the cup of tea I had with Jock, cooked over hot slag at 4.00 A.M.

A battle certainly, to make an ingot — trench work, in a quiet sector, perhaps, but a year-after-year affair. The multiform steel prop which civilization hung upon came to me for a moment — rails, skyscrapers the locomotive just passed, machinery that was making the ornament and substance of the environment of men. It rested on muscle and the will to push through "long turns," I thought. It could slip so easily. A huge mistaken calculation: not enough coal or cars to carry it. Or what if the habitual movements of the muscles were broken, or the will fallen into distemper? Suppose men thought it not worth the candle, and stopped to look on?

Were we to get more of the kind of civilization we knew, conquer more ground, or have less of it? It depended on the battle. And that hung, I was sure, on the morale of the fighter. I wondered if it was n't cracking badly —

But at this point I considered how late it was, and whether it was not time for bed, that I might not have bad morale myself, with a headache added to it, at 6.00 A.M.

The roar again — I began breaking it up once more into the fragments of grind and rattle that composed it. In imagination I jumped on the step of the charging-machine as it moved on its rails past Seven. It shook and jarred grumpily about its business, I thought.

Near Five I got off, and started to make front-wall. I remembered how I felt on a front-wall a few weeks ago. I had tried to throw my mind into the unsleeping numbness that protects a little against the load of monotony. Other men I had seen do it, drawing a curtain over nine tenths of their brain; not thinking, but only day-dreaming faintly behind the curtain, leaving enough attention to the fore for plunging the shovel into dolomite, and keeping the arms out of heat.

Other passages from open-hearth shifts came into my mind in violent contrast. Shorty, who was always clearly to be distinguished anywhere on the floor because he wore his khaki shirt outside his pants, quarreled with me one day, and showed his temper, as one shows temper in Italy. He stood by the drinking fountain back of Number 4, hair on end, chest bare, his eyes a little bloodshot, and his mouth sullen and drawn at the corners, as it always was. The argument was about a shovel. Shorty took out a long knife from his pocket and explained its use in argument.

I remembered how the mill stayed in your mind when you left it. In the hour or so in which you washed up, walked home, ate, and went to bed, it loomed as a black sheet-iron foreman, demanding that you get to bed and prepare for the noise and jar it had in store for you at 5.00 o'clock. That sense of imminence was a thing to bear, especially if you wondered whether sleep would come at all.

Then there were long strings of neutral days when you did not think well of life, or ill of it. And there were the occasional satisfactions. The keen pleasure of acquiring a knack, as when I learned to "get it across" in back-wall. And the pleasures of rough-house. Jock, the first-helper on Seven, had once told me in a burst of enthusiasm for furnace work that he "liked the game because there was so much hell-raisin' in it."

In the midst of listening to the roar, and thinking of shifts, good and bad, it occurred to me abruptly that men would make front-walls in front of hot furnaces for several hundred years, in all likelihood. I wondered. Perhaps Mr. Wells's army of inventors would alter that. For several hundred years, thousands of men had labored without imagination or hope in

I TAKE A DAY OFF

Egypt, and built the Pyramids. There were similarities. Civilization rested on the uninspired, unimaginative drudgery of nine tenths of mankind. "There have always been hewers of wood, and drawers of water," I heard some elderly person say at me, in a voice of finality.

I did not stop to reply to the implications of that sentence in my own mind, but thought more closely of the Pyramid-builders I had known in the pit.

Marco drew Croatian words for me with a piece of chalk on his shovel, and I put down English ones for him. He had attended night school after working twelve hours a day in Pittsburgh. But Marco was, perhaps, exceptionally gifted.

The jobs we did were pick-and-shovel jobs. But have you ever used a pick on hot slag? There is judgment and knack, and he is a fool who says that "anyone can do the job." Whenever the chance for special skill happened by, as in hooking the crane to a difficult piece of scrap, there was all abundance, and much rivalry to show it off. Could such substance of "knacks" ever grow into anything more for this "nine tenths of mankind?" I wonder.

How much of strength, of skill, of possible loyalty, does modern industry tap from the average Hunky?

I asked the question, but did not answer it — for modern industry. I answered it for the gang in the pit and the crew on the stoves of the blast-furnace.

Not half.

There were vast unused areas of men's minds and of their muscles, as well as of their powers of will that were wholly unreached in the rough job adjustment of modern industry. I mean among the so-called groups of "lower intelligence." It was all interesting speculation whether any engineer would ever find a means of tapping this unused voltage.

I suddenly thought how inconceivable the stoppage of that roar would be. A silent valley, with all those ordered but gigantic forces stopped, would be almost terrible. But just such a silence was likely to happen. By a walk-out.

The great strike had been going a week, in other towns — tying up the steel production of the country. Meetings had followed, and riots, with an occasional bloody conflict with the "mud guard" of Pennsylvania.

Part of that untapped force! I said to myself—dynamos of power of all sorts. Would it bludgeon over a change in steel conditions, or flow back, waste voltage, into the ground?

The rumble in the valley again. Could I hear the shake of the charging-machine at this distance? The Bessemer glow had changed. The nail mill roar seemed to increase.

I went down the hill. When I reached Mrs. Farrell's and climbed into my back room, I set the alarm for 4.00 o'clock, putting the clock a foot and a half from the bed. It has a knob on top, and you can stop it by knocking down the knob with the palm of your hand. I went to sleep, to dream about the men who built the Pyramids.

IX

"NO CAN LIVE"

I WENT into the employment office one day, to fix up the papers of my transfer to the blast-furnace, and got into a talk with Burke, the employment manager, about personnel work.

"What do you think of the game?" I asked.

"It's great," he returned; "it's working with human material — that's what it is; there's nothing like it.

"But," he added, "if you have any ideas about unions keep them in the back of your head — that is, if you want a job in steel. They won't stand for that sort of thing."

He looked down on his desk, where there was a news-clipping of the demands of the American Federation of Labor's Strike Committee — the twelve demands. He pointed to it.

"We give them practically all of these here in Bouton," he said, "all but two or three."

"The eight-hour day?" I queried.

"Yes, we give them the eight-hour day. Overtime for everything over eight hours."

"Could I stop work to-day after eight hours' work on the furnace?" I asked "Could anyone before six o'clock, and hold his job?"

"Oh, no," he returned.

"I should call that a twelve-hour day," I said.

The "safety man" came in, and interrupted. He was a stocky young man with the intelligent face of an engineer.

"That man might do something for the steelworker," I thought.

The men on the furnaces were talking about the strike that day. One young American said: "Well, strike starts Monday. Damned if I won' go if the rest do."

There were no leaders about, and it was unlikely, perhaps, that any would appear. There seemed to be a current opinion that any organizers "get taken off the train before they get to Bouton."

The Old Home Week Carnival had been called off through the influence of the mill authorities. They were afraid of a strike committee coming from the next town, and having a parade to lead the men out.

A special train went through Bouton that day at about five o'clock. Everyone watched it from the furnaces, and speculated what it meant. It was a double-header and passed through at top speed.

"Troops going to quell strike riots," the Assistant Superintendent, Lonergan, suggested. "A lot of those fellers are overseas men of the National Guard. They 're havin' trouble with 'em. I don't blame the boys a damn bit for not wantin' 'to preserve order in the steel towns,' as the papers call it," he concluded, with a grin.

Haverly, an American blower, came up. "Fight for democracy overseas and against it over here," he said.

It is difficult to say what the men would have done if they had leadership. They had none, since no organizers whatever appeared, and no speech-making occurred in town. There was pretty good feeling toward the company itself, which is, I believe, one of the best. A deep-seated hatred, however, existed against the whole system of steel. There was anger and resentment that ran straight through, from the cinder-snapper to the high-paid blowers, melters, and in some cases, to the superintendents.

I was quite amazed — because of what the newspapers were continually saying — at the absence of any sociological ideas whatever. I remember one day I met my first and only Socialist. He was a stove-tender of great skill and long experience; he told me how bad he thought war was, and how he could n't understand why people did n't live in peace and be sociable with one another. But, though there were few doctrines, except in rare instances, there was a mighty stream of complaint against certain things such as the company-owned town, the twelve-hour day, the twenty-four-hour shift, the seven-day week, and certainly remediable dangers. It pervaded all ranks.

There were certain days in my summer in the mills that burned among the others like a hot ingot of steel on the night-shift. One of them was the cleaning out of No. 15 stove early in my gang apprenticeship. Ordinarily, the duties of the stove gang were to move leisurely from stove to stove while they were alight, and remove cinder from the combustion chambers. It was pried up with a crowbar, and hoed out on to a wheelbarrow. But when a stove was cooled for thorough cleaning, we did our real work.

The gas was turned off in the combustion chamber on the night-shift, and the stove allowed to cool for several hours. We prepared to go inside her, the next morning, to cutaway the hardened cinder. John, the Slav, went in first, with pick and shovel, and worked an hour. Then Tony turned to me.

"You go in with me, I show you," he said.

We put on wooden sandals, foot-shaped blocks an inch thick, with lacing straps, donned jackets that buttoned very tight in the neck, and pulled down the ear-flaps of our kersey caps. Over our eyes we wore close-fitting goggles. We looked like Dutch peasants dressed for motoring. The combustion chamber is a space eight or ten feet long by three or four wide. It was partly filled with cooling cinder, some of it yielding to the pick, some only to the bar and sledge. Someone shoved an electric light through the hot-blast valve, and the appearance of the place was like a mine gallery. The chamber was hot and gaseous, but it was quite possible to work inside over an hour. After Tony had loosened several shovelfuls, I could see that the pick failed against a great shelf of the stuff that glowed red along its base.

"Bar," he called.

The bar came in through the little round door in three or four minutes. He held it for me, and I sledged. It needed a little work like this to make you yearn for real air. The heat weakened you quickly. We worked about forty minutes, and then lay on our bellies and wriggled out. The means of entrance and egress is a small door, about fourteen inches in diameter, which means absorbing a good deal of cinder when you caterpillar through.

We finished the whole job in three hours, and then went to the other side of the stove and cleared out half a carload of flue-dust from the brick arches that compose the groundwork of that side of the stove. The dust lay a foot or two thick, and one man worked with a shovel in each archway. Here it was hardly hot at all, but merely thick with the red iron-dust. As you bent over inside the archways, knee-deep in the stuff, it would rise and settle on your arms and shoulders; you kept up a blowing with your nose to keep it out. Some of it was hard and soggy, and pleasanter shoveling. Five or six of us could work inside the stove at once, in the different archways, each with a teapot lamp near by, and a large, light shovel. Men at the entrances hoed the stuff out as we threw back.

But it was the next day's cleaning that I remember most strongly. The word went about that we were to "poke her out," to-morrow. That night the gang, and especially John, the Italian, instructed me very seriously to bring a selected list of clothing the next morning: a jacket, a cap with flaps for the ears, two pairs of gloves, and two bandanna handkerchiefs.

"NO CAN LIVE"

We went on top of Number 15, and started to dress for the job of poking her out. Over our faces we tied the handkerchiefs, leaving only our eyes exposed. Our necks and ears were covered with the winter caps, our hands with two pairs of gloves.

The stove, as I said, looked like a very tall boiler: half was a long brick-lined flue, where the gas burned; half, a mass of brick checkerwork for retaining the heat. Masses of flue-dust had clogged the holes in the checkerwork and reduced its power for holding heat. It was our job to poke out that dust.

A ladder leading from a trapdoor in the side of a stove at the Aliquippa plant, just as Walker describes it. Aliquippa Works, Jones and Laughlin Steel Company. Photo: Don Inman Collection, Beaver County Industrial Museum, Geneva College, Beaver Falls, PA.

John and Mike and I unscrewed the trap at the top very deliberately, and dropped a ladder down. There was a space left at the top of the checkerwork for cleaning purposes. We worked on top of that.

Jimmy, I think, went in first, taking a teapot lamp with him and a rod. In three minutes he was out again, and Mike down. I began to wonder what the devil they faced for three minutes in the chamber. Tony looked at me and said, "I teach you, now."

I tied the handkerchiefs around my face, sticking the end of one in my collar, and followed Tony.

My first sensation, as I stepped off the ladder to the checkerwork inside the stove, was relief. It was hot, but quite bearable. I picked my way slowly to Tony, and tried to study in the dull light his motions with the rod. The dust was too thick and the lamp guttered too violently for me to follow his hand. I bent over to watch the end of his rod, and recoiled. I felt as I had when the ladle got under me on the manganese platform — flame seemed to go in with breath. It was the hot blast that continued to rise from the checkerwork, and made it impossible to work beyond three minutes in the stove.

When I mounted the ladder, and moved out into the air, I thought, "I haven't learned much from Tony, except that he somehow cleaned the checkerwork, and it's best to keep the head high; no more bending."

Five minutes passed, and I was scheduled to take my turn alone. Every man poked three holes and came up. I was full of resolutions for glory and poked four, coming up rather elated. John looked at me sadly when I stepped off the ladder.

"What's the matter, Charlie? You only poke 'em half out." He simulated my motions with the rod. I had n't qualified.

John, the Slav, was tying his handkerchief back of his ears.

"I show him; you come with me, Charlie, I show you all right."

I wasn't gleeful. The last time I had done a job with John, we had carried pipes, many more at a time than anyone else. John, I anticipated, would stay in the stove, poking away, till ordinary mortals lost their lungs.

He picked up a poking rod, after very carefully putting on his gloves, and went over to the ladder, descending slowly. I followed him with my teeth in my lips, feeling for the rungs of the ladder with my feet, and holding my poking rod in my right hand. When I stepped off at the bottom, I felt my fingers closing over the bent handle of the rod in a death grip. I determined on no half-way poking.

View of the interior of a stove. We see the checkers (small holes), similar to those Walker describes, on the right. The large hole to the left is the combustion chamber, much like the one into which a man fell in the story. Aliquippa Works, Jones and Laughlin Steel Company. Photo: Don Inman Collection, Beaver County Industrial Museum, Geneva College, Beaver Falls, PA.

John set to work at once, and I after him, rattling my rod in the checkerwork with all my strength, and pushing her in up to the hilt. I did three holes, and John four. My lungs were like paper on fire, when John turned to go up. We climbed out of the hole and took down the handkerchiefs. The gang looked at me, and then at John.

"He do all right," he cried rather loudly, "every time all right."

I felt extraordinarily elated, and much as if John had given me a diploma, with a cum laude inscribed in gold letters.

There was later a trip down inside with Jimmy. He shouted a great many things at me in Anglo-Italian, which caused me a good deal of anxiety but no understanding. I learned on coming up that he was trying to tell me not to approach the combustion chamber adjoining the checkerwork. That is a clear shaft to the bottom. I was given in some detail the story of the man who fell down a year ago, and was found with no life in him at the bottom.

"Kill him quick," said John the Italian; "take him out through hot-blast valve."

Two burns on my wrists were an embarrassing legacy of this affair, for they required an explanation whenever I took off my coat. My arms were too long and shot from my sleeves, when poking out, and got exposed to the gas and flame, which were still rising in the checkerwork.

This incident put me into good standing with John, the Slav, I am delighted to say. He was a stoical person, Without much conversational warmth, but he approached me at the foot of the furnace steps in the late afternoon; "Some people, no show new man; I show him, I Slovene, no Italian, been in this country eighteen year." That was about all, but enough for a basis of friendship.

I sat on my bed and sewed up a rip in my trousers, eleven inches long. It was lucky I had salvaged that khaki "housewife" from the army. My gray flannel shirt lay on the bed. There were little holes you could pass matches through, all over it, with brown edges that sparks had made.

Would that sleeve last?

I made it last.

Then there were the pants.

That second-hand paint-spattered pair of mine had lasted five days. The next, a sort of overally kind, had stood it a month, the last week in entire disgrace; these mohair ones I got at the Company store were going yet. But the seat needed emergency attention.

After sewing-time, I got up and stared out of the window at Mrs. Farrell's four stalks of corn. They were doing well. I looked across at the back road, along which a junk-dealer's wagon jangled. The mud cliff was the horizon of the prospect. I watched a little stream going down it among roots, which I had watched a good many times before, and finally picked up my army field-shoes, and took them out to a Greek cobbler for resoling.

I shall remember for all time the "blowing in" of Number 9, which means its first lighting up. A blast-furnace, once lit, remains burning till the end of its existence. I got inside her, and was delighted to satisfy a deep-seated curiosity: we crawled in the cinder notch. The hearth of the furnace lay six or eight feet below the brick flooring, and the effect of standing inside, with the fourteen round blowpipe holes admitting a little sunlight, was like being in a round ship's cabin, with fourteen portholes, except that the hollow furnace shot up to dark distances that the light did n't penetrate.

We built a scaffolding six or eight feet above the hearth to hold firewood, and filled in beneath with shavings and kindling. Then we took in cords upon cords of six-foot sticks and set them on end on top; there were two or three layers of these, and on top of them, according to the orthodox rule, were dumped quantities of coke, dumped down from the top, of course, by skips; and above that, light charges of ore. Below the scaffold, we spent half a day arranging kindling, with shavings placed at each blowpipe hole. When the wood was arranged, — a three-days' job, — the crane brought us some barrels of petrol, and we poured about half a one in each blowpipe hole. The cinder notch was likewise thoroughly provided with soaked shavings. That was to be the torch.

Men assembled as at a house-raising. Nobody worked from 11.00 to 12.00 on the day of blowing in Number 9. From all parts of the blast-furnace they came and arranged themselves about the cinder notch, and on the girders above. The men and their bosses came. There was the labor foreman, and the foreman of all the carpenters, of all the window-glass fixers, all the blowers, the electricians, the master mechanic. Then came the superintendent of the open-hearth and Bessemer, Mr. Towers, and Mr. Brown his boss; and, finally, Mr. Erkeimer, the G. M., with an unknown Mr. Clark from Pittsburgh.

We waited from 11.00 to 12.00 for Mr. Clark to come and drop a spark into the shavings. When he arrived the crowd parted quickly for him, and, with Mr. Erkeimer and Mr. Swenson, he stood talking and smiling for some minutes more at the notch. Mr. Clark was a tall slender person, with glasses and an aspect of unfamiliarity with a blast-furnace environment. No one knew, or ever found out, who he was. Mr. Swenson showed him, very carefully, how to ignite the shavings with a teapot lamp. Twice the photographer, who had come early, got focused for the awful moment, and twice Mr. Clark deferred lighting the shavings and went on talking with Mr. Swenson. Finally, he bent over and lit them. Mr. Swenson rapidly turned to the gang behind him.

"Three cheers for Mr Clark!" he cried, raising his hand. When it is recalled that none of us knew the man we cheered, it was n't a bad noise. The furnace smoked lustily in a few minutes, and several helpers rushed around it to thrust red-hot tapping bars in the blow-holes. They ignited at once the petroleum and shavings packed around them.

Immediately after the cheers, Mr. Swenson's bright looking office-boy hurried through the gang with a box of cigars, another immemorial custom in operation. The more aggressive got cigars, then disappeared. It was a little odd during the afternoon to see a sweat-drenched cinder-snapper at his work with a long black cigar between his teeth. When they were burned out, the department settled back to normal production.

Many years might pass before such another occasion in that place. During that period there would be no slackening of the melting fires, or of the work of the helpers who kept them alive.

I stood on the platform waiting for the 10.05 train, and turned for a look at the landscape of brick and iron. I remembered a Hunky who had worked in the tube-mill for eighteen years and at length decided to go back to the old country. On the day he left, he went out the usual gate at the tempered after-work pace, walked the gravel path to the railroad embankment, and stopped for a moment to look back at the mill. He stood like a stone-pile on the embankment for a quarter of an hour, looking at the cluster of steel buildings and stacks. He had spent a life in them, making pipe, and I have n't a doubt this was the first time it came to him in perspective. From my own brief memories, I could guess at those fifteen minutes: pain, struggle, monotony, rough-house, laughter, endurance, but principally toil without imagination.

I thought quickly over my summer in the mills, and it looked rather pleasurable in retrospect. Things do. There 's a verse on that sentiment in Lucretius, I think. I thought of sizzling nights; of bosses, friendly and unfriendly; of hot back-walls, and a good first-helper; of fighting twenty-four hour turns; of interesting days as hot-blast man; of dreaded five-o'clock risings, and quiet satisfying suppers; of what men thought, and did n't think — And again, of how much the life was incident to a flinty-hearted universe and how much to the stupidity of men. I knew there were scores of matters arranging themselves in well-ordered data and conclusion in my head. I had a cool sense that, when they came out of the thinking, they would not be counsels of perfection, or denunciations, but would have substance, be able to weather theorists, both the hard-boiled and the sentimental, being compounded of good ingredients — tools, and iron ore, and the experience of workmen.

Is there any one thing though that stands out? I heard the train whistle a warning of its arrival. Perhaps, if a very complicated matter like the steel-life can be compounded in a phrase, it had been done by the third-helper on Six. On the day we had thrown manganese into a boiling ladle, in a temperature of 130°, he had turned to me slowly and summed it all up.

"To hell with the money," he said; "no can live!"

Epilogue

A FURNACE-WORKER
TALKS OVER THE TWELVE-HOUR DAY

EPILOGUE

I HAVE tried to put down the record of the whole of my life, as I lived it, and the whole of my environment, as I saw and felt it, among the steel-workers in 1919. To me the book is the story of certain obscure personalities, and the record of certain crude and vital experiences we passed through together. I think it may be read as a story of men and things.

Many people, however, have asked me the questions: What were the conditions in steel and what is your opinion of them? What do you think of the twelve-hour day? or, How bad was the heat? and the like. And, What do you suggest? Since no man who has worked in an American steel mill, whatever his sympathies or his indifference, can fail to have opinions on these points, I have decided to set down mine, for what they are worth, as simply and informally as I can.

There is a proper apology, I think, that can be made for the presumption of conclusions based upon an individual experience. An intimate and detailed record of processes and methods and the physical and mental environment of the workers in any basic industry is rare enough, I believe, except when it is heightened or foreshortened for a political purpose. No industrial reform can rest upon a single narrative of personal experience; but such a narrative, if genuine, can supply its portion of data, and possibly point where scientific research or public action can follow.

Let me state my bias in the matter as well as I can. I was by no means indifferent to economic and social values when I began my job; in fact, I confess to being interested keenly in most of them. But I never sought information as an "investigator." Most of my energy of mind and body was spent upon doing the job in hand; and what impressions I received came unsought in the course of a day's work. I began my job with an almost equal interest in the process of steel-making, the administration of business, and the problem of industrial relations.

Some apology I owe to the several hundred steel-workers with whom I worked, and the many thousands in other mills, since most of them know from a far longer and deeper experience the conditions and policies of which I speak. My sole reason for raising my own voice in the presence of this multitude of authorities is that the Hunkies, who constitute the major part, are unable either to find an audience or to be understood if they find one. Again, they are like Pete, who, when I asked him what were the duties of a third-helper, which I have described to the length of several pages in this book, replied: "He has a hell of a lot to do." And as to the American workers and bosses, most of them lack the opportunity of any speaking that will be heard beyond their own furnaces; and, again, they are too close to their environment to see what is in it. They are natives, while I am more nearly a foreigner with something of the freshness and perspective that a foreigner brings.

I want to add that the management of the mill where I worked was a body of men exceedingly efficient and fair-minded, it appeared to me; and any remarks upon the twelve-hour day, or other conditions, are critical of an arrangement typical of American steel-management as a whole, and not of individuals or a locality.

The twelve-hour day makes the life of the steel-worker different in a far-reaching manner from the life of the majority of his fellow workers.

It makes the industry different in its fundamental organization and temper from an eight-hour or a ten-hour industry.

It transforms the community where men live whose day is twelve hours long.

"What is it really like? How much of the time do you actually work? Are you 'all in' when you wash up in the morning after the shift, and go home?"

EPILOGUE

 To tell it exactly, if I can: You go into the mill, a little before six, and get into your mill clothes. There may be the call for a front-wall while you're buttoning your shirt. You pick up a shovel and run into a spell of fairly hot work for three quarters of an hour. On another day you may loaf for fifteen minutes before anything starts. After front-wall, you take a drink from the water fountain behind your furnace, and wash your arms, which have got burned a little, and your face, in a trough of water. A "clean-up" job follows in front of the furnace, which means shoveling slag — still hot — down the slag-hole for ten minutes, and loading cold pieces of scrap, which have fallen on the floor, into a box. Pieces weigh twenty, forty, one hundred pounds; anything over, you hook up with a chain and let the overhead crane move it. This for a half-hour.

 Suddenly someone says, "Back-wall!" Lasts say thirty or forty minutes. It's hot — temperature, 150° or 160° when you throw your shovelful in — and lively work for back and legs. Everybody douses his face and hands with water to cool off, and sits down for twenty minutes. Making back-wall has affinities with stoking, only it's hotter while it lasts. The day is made up of jobs like these — shoveling manganese at tap-time, "making bottom," bringing up mud and dolomite in wheelbarrows for fixing the spout, hauling fallen bricks out of the furnace.

 They vary in arduousness: all would be marked "heavy work" in a job specification. They are all "hard-handed" jobs, and some of them done in high heat. Between, run intervals from a few minutes to two or three hours. From some of the jobs it is imperative to catch your breath for a spell. Sledging a hard spout, making a hot back-wall, knocks a gang out temporarily — for fifteen or twenty minutes; no man could do those things steadily without interruption. It is like the crew resting on their oars after a sprint. Again, some of the spells between are just leisure; the furnace does n't need attention, that's all; you're on guard, waiting for action. Furnace work has similarities with cooking; any cook tends his stove part of the time by watching to see that nothing burns up.

I have had two or three hours sleep on a "good" night-shift; two or three "easy" days will follow one another. Then there will come steady labor for nearly the whole fourteen hours, for a week.

So, briefly, you don't work every minute of those twelve hours. Besides the delays that arise out of the necessities of furnace work, men automatically scale down their pace when they know there are twelve or fourteen hours ahead of them: seven or eight hours of actual swinging of sledge or shovel. But some of the extra time is utterly necessary for immediate recuperation after a heavy job or a hot one. And none of the spells, it should be noticed, are "your own time." You're under strain for twelve hours. Nerves and will are the Company's the whole shift — whether the muscles in your hands and feet move or are still. And the existence of the long day makes possible unrelieved labor, hard and hot, the whole turn of fourteen hours, if there is need for it.

Inseparable from the twelve-hour day in the open-hearth where I worked were the twenty-four-hour shift, and the seven-day week.

What does it mean to make steel twenty-four hours a day? to your muscles, to your thoughts, to the production of steel? Sunday morning, at 7.00, you begin work. There is an hour off at 5.00 P.M. Front-wall, fix spout, tap, back-wall, front-wall, fix spout, tap, back-wall — the second half is something of a game between time and fatigue. For a hot back-wall, or sledging out a bad tap-hole, may as easily come upon you at 5.00 or 6.00 of the second morning as at noon of the first day. I've worked "long turns" that I did n't mind overmuch, and others that ground my soul. If you are young and fit, you can work a steady twenty-four hours at a hot and heavy job and "get away." But in my judgment even the strongest of the Czechoslovaks, Serbs, and Croats who work the American steel-furnaces cannot keep it up, twice a month, year after year, without substantial physical injury. "A man got to watch himself, this job, tear himself down," the second-helper on Seven told me. He had worked at it six years, and was feeling the effects in nerves and weight. Let me make an exception: one Hunky, a helper on Number 4, was famed for having "a back like a mule." He could, I am sure, work seven twenty-four-hour shifts *a week* with comfort. But for all other men, with the exception of Joe, the long turn is an unreasonable overtaxing of human strength. Lastly, the effort of will, the "nerve" that the thing calls for in the last hours before that second morning, is too heavy a demand, for any wages whatever. The third-helper on Number 8 took, I think, a reasonable attitude when he said: "To hell with the money, no can live!"

The "long turn" leaves a man thoroughly tired, "shot," for several shifts following. As I said in the first part of this book, it is hardly before Friday that the gang makes up sleep and comes into the mill in normal temper. Here is the condition. You have ten hours for recuperation after twenty-four hours' work. Washing up in a hurry, getting breakfast, and walking home gets you in bed by 8.00. Eight hours' sleep is the best you can get. At 4.00 o'clock you must dress, eat, and walk to the mill. Men who live an hour or more from the mill, as some do, must, of course, subtract that time as well from sleep. After the ten hours off, you return to the mill at 5.00, to begin another fourteen-hours' steel-making. That night is unquestionably the worst of the two-weeks cycle. The nervous excitement that helps any man through the twenty-four turn has gone-quite.

The seven or eight hours of day sleep seem to have taken that away without substituting rest; and what you have on your hands is an overfatigued body, refusing to be goaded further. My observation was that, on this Monday after, men made mistakes; there were arguments, bad temper, and fights, and a much higher frequency of collision with the foreman. Efficiency, quality, discipline dropped.

The other accompaniment of the twelve-hour shift, the averaging of seven working-days per week, has, I am convinced, an equally bad physiological effect upon the healthiest of men. As I have said earlier, "the twenty-four hours off," which comes once a fortnight on alternate weeks to the twenty-four-hour shift, is a curiously contracted holiday. It comes at the conclusion of fourteen hours' work on the night-shift and is immediately followed by ten hours' work on the day-shift. As far as I could observe, men went on a long debauch for twenty-four hours, or, if the week had been particularly heavy, slept the entire twenty-four. In the first instance they deprived themselves of any sleep, and went to work Monday in an extraordinarily jaded condition. In the second, they forfeited their only holiday for two weeks.

Another feature that impresses you when you actually work under the system is that the sleep you get is troubled, at best. You are compelled to go to bed one week by day, and the next by night. By about Friday, I found my body getting itself adjusted to day sleep; but the change, of course, was due again Monday. And yet, by comparing my sleeping hours with those of my fellow workers, I found my day rest was averaging better than theirs. Many of them, I found, went to bed at 9.00 in the morning and got up about 2.00. They complained of being unable to sleep properly by day. The body will adjust itself to continued day sleeping, I know; but apparently not to the weekly shifts, from day sleep to night sleep, customary in steel.

EPILOGUE

The "long turn" of twenty-four hours and, the "seven-day week" I have never heard defended, either in the mill by any foreman or workman, or outside by any member of the management, or even in a public statement. If, by an arrangement of extra workers, it were possible to eliminate these features and still keep the twelve-hour work-day for six days a week, there would, I think, be a certain number of men ready enough to work under that arrangement. I met one man, for example, who said: "Good job, work all time, no spend, good job save." There are a few foreign workers whose plan is to work steadily for ten or fifteen years, and then carry the money back to the old country. These men are willing to spend the maximum time within mill walls, since they have no intention of marrying, settling down, and becoming Americans. But their numbers are small, and the desirability of their type is questionable. It is unwise, at any rate, to build the labor policy of a great industry in their interest.

On those first night-shifts I wondered if my feelings on the arrangement of hours were not solely those of a sensitive novice. I'd "get used to it," perhaps. But I found that first-helpers, melters, foremen, "old timers" and "Company men" were for the most part against the long day. They were all looking forward, with varying degrees of hope, to the time when the daily toll of hours would be reduced.

The twelve-hour day gives a special character to the industry itself as well as to the men. I remember noticing the difference in pace, in tempo, from that of a machine shop or a cotton mill. Men learn to cultivate deliberate movement, with a view to the fourteen-hour stretch they have before them. When I began work with a pickaxe on some hot slag, on my first night, I was reproached at once: "Tak' it eas', lotza time before seven o'clock." And the foremen fell in with the men. They winked at sleeping, for they did it themselves.

Another kind of inefficiency that flowed quite naturally from excessive hours was "absenteeism," and a high "turnover" of labor. Men kept at the job as long as they could stick it, and then relaxed into a two or three weeks' drunk. Or they quit the Company and moved to another mill, for the sake of change and a break in the drudgery. I remember an Austrian with whom I worked in the "pit," who said he was going to get drunk in Pittsburgh, go to the movies, and move to Johnstown the following Monday. He had been on the job three weeks. New faces appeared on the gangs constantly, and dropped out as quickly. I achieved my promotion from common labor in the pit to the floor of the furnace by supplying on a twenty-four-hour shift, when absentees are apt to be numerous, and it is hard fully to man the furnaces. The company kept a large number of extra men on its pay roll because of the number of absentees, and the turnover percentage ran high.

It is impossible to live under this loose régime — with high turnover, and the work-pace necessarily keyed low because of the excessive burden of hours spent under the roof of the mill — and not wonder if there is n't an engineering problem in it. The impression was of a vast wastage of man-hours. The question suggested itself: "Is it in the long run, good business — an efficient thing?" An exhaustive investigation by engineers and economists could surely be made to answer this question.

People ask: "Is there any mechanical or metallurgical reason for the twelve-hour day?" The answer is: No. There are several plants of independent steel companies that run on a three-shift, eight-hour basis; and the steel mills in England, France, Germany, and Italy operate with three eight-hour shifts. The long day is not a metallurgical *necessity*, therefore. The metallurgical *explanation* of the twelve-hour day, however, is this. The process of making iron or steel is necessarily a continuous one, because the heat of the furnaces must be

EPILOGUE

conserved by keeping up the fires twenty-four hours a day. So the division into either two shifts of twelve hours or three shifts of eight becomes imperative. Other industries might reduce their hours gradually from twelve to ten, and then to nine. With steel the full jump from twelve to eight must be made. Without doubt, this metallurgical factor accounts in some measure for the conservatism of the steel companies in making the change.

It is none of my business, in summing up a personal experience, to review the story of steel mills which have undertaken a three-shift plan of operation, of eight hours each, in place of the two shifts of twelve. But the study has been made by engineers and economists, who have collected figures as to the cost of operation on an eight-hour basis as contrasted with a twelve. The increased cost in product which such a change would entail is between three and five per cent.[1]

The community of workers takes on a special character, where men live whose day is twelve hours long. "We have n't any Sundays," the men said; and "There is n't time enough at home." This is the most far-reaching effect of "hours" in steel, I think, and easily transcends the others.

"What do you do when you leave the mill?" people ask. "On my night-week," I answer, "I wash up, go home, eat, and go to bed." Anything that happens in your home or city that week is blotted out, as if it occurred upon a distant continent; for every hour of the twenty-four is accountable, in sleep, work, or food, for seven days; unless a man prefers, as he often does, to cheat his sleep-time and have his shoes tapped, or take a drink with a friend.

The day-week is decidedly better. You work only ten hours, from seven to five. Those evenings men spend with their families,

---

[1] The Three Shift System in Steel—Horace B. Drury: an address to the Taylor Society and certain sections of the Am. Soc. Mech. Engineers and of the Am. Inst. Electr. Engineers, Dec. 3, 1920.

or at the movies, or going to bed early to rest up for the "long turn." It is not, however, as if it were a "ten-hour industry." Some of the wear and tear of the seven fourteen-hour shifts of the night-week protracts itself into the day-week, and you hear men saying: "This ten-hour day seems to tire me more than the fourteen; funny thing." However the week may be divided up, it is impossible to keep the human body from recording the fact that it averages seven twelve-hour days, or eighty-four hours of work, in the week.

For the men who did a straight twelve hours, "six to six" for seven days, the sense of "no time off" was very strong. I worked these hours for a time on the blast-furnace, and remember that the complaint was, not so much that there wasn't some bit of an evening before you, but that there was no *untired* time when you were good for anything — work or play. When you had sat about for perhaps an hour after supper, you recovered enough to crave recreation. A movie was the very peak to which you could stir yourself. There were men who went further. I knew a young Croat in Pittsburgh who attended night-school after a twelve-hour day. But he is the only one of all the steel-workers I met who attempted such heroism. And he had to stop after a few weeks.

Now it should be mentioned that some of the social life that most workers find outside the mill gets squeezed somehow into it. In the spells between front-walls we used to talk everything, from scandal about the foreman to the presidential election. The daily news, labor troubles, the late war, the second-helper's queer ways passed back and forth when you washed up, or ate out of your bucket, or paused between stunts. Then there was kidding, comradely boxing, and such playfulness as hitching the crane-hooks to a man's belt. One first-helper remarked: "I like the game because there's so much hell-raisin' in it."

EPILOGUE

But this is hardly a substitute for a man's time to himself, for seeing his wife, knowing his own children, and participating in the life of larger groups. Soldiers have a faculty for taking so good-humoredly the worst rigors of a campaign, that some people have made the mistake of turning their admirable adaptability into a justification for war.

The twelve-hour day, I believe, tends to discourage a man from marrying and setttling into a regular home life. Men complained that they did n't see their wives, or get to know their children, since the schedule of hours shrunk matters at home to food, sleep, and the necessities. "My wife is always after me to leave this game," Jock used to say, the first-helper on Seven. Mathematically, it figures something like this: twelve hours of work, an hour going to and from the mill, an hour for eating, eight hours of sleep — which leaves two hours for all the rest, shaving, mowing the lawn, and the "civilizing influence of children."

I have no brief to offer for the eight-hour day as a general panacea for evils in industry. I merely bear witness to the fact that the twelve-hour day, as I observed it, tended either to destroy, or to make unreasonably difficult, that normal recreation and participation in the doings of the family group, the church, or the community, which we ordinarily suppose is reasonable and part of the American inheritance.

Steel has often been described by its old timers as a "he-man's game." That has even figured as an argument against any innovation that might lighten the load of the workers in it, and against any change in the twelve-hour day itself. The industry has certainly a rough-and-tumble quality and a dangerous streak in it, that will always call for men with some toughness of fibre. But I question whether the quality of the men it attracts, and the type it moulds within its own ranks, will ever be improved by the twelve-hour day. The excessive hours, I know, operate as a check against many younger men, who would otherwise enter the industry. The inherent fascination of making steel is, I think, very great. It was for me. But the appeal is the mechanical achievement of the industry, its size, power, and importance, even its dangers. The twelve-hour day, on the other hand, tends to place a premium on time-serving and drudgery, in lieu of the more masculine qualities of adventure and initiative.

Afterword
WHERE IS BOUTON, PENNSYLVANIA?

Where is Bouton, Pennsylvania? If you check a modern map or one from the period about which Walker writes, you will find no town in Pennsylvania with that name. Yet, this Bouton is not a mythical town. The mill, the people, the work that they did, and the town were all real.

Why would Walker choose to conceal the real name of the place or the company at which he toiled? As he has been dead for a number of years, the true answer to that question will never be known. However, while he claims that he "never sought information as an investigator," his interest was in the socio-economic issues of the times, the twelve-hour day and the "long turn" in particular, which were indeed controversial. Some inferences can be made that the plant managers may have known of Walker's ideas and still let him work at the plant. He very likely wanted to protect those people who gave him the opportunity to learn and write about the issues.

The 24-hour or so-called "long turn" was soundly criticized by many social workers and humanitarians of the time. European steel companies had long since dispensed with this practice. Other industries in America had done so as well, but steel is inherently a twenty-four-hour-a-day, seven-day-a-week operation and was difficult to change without a great deal of disruption. The safety and health conditions under which these men labored was another area requiring vast improvement. Finally, the ethnic and racial discrimination practiced seems to be something that the author wished to highlight, but he wouldn't have wanted to embarrass his newfound "friends." To quote Walker, "My sole reason for raising my own voice in the presence of this multitude of authorities is that the Hunkies, who constitute the major part, are unable either to find an audience or to be understood if they find one." Certainly, all of these conditions needed improvement.

Given the above, it is easy to believe that Walker would want to shield his sources, just as any writer might do today. But he wasn't as successful as he may have wished, for throughout the book he leaves a trail of clues to the true identity of "Bouton."

On the very first page of Walker's Foreword we find that the mill in question is near Pittsburgh. In almost any other location this would probably be an enormous hint. But in Pittsburgh, the largest steel center in the world, it amounts to a needle in a haystack. The only other reference to a real site near Pittsburgh is found on page 118 when Walker talks with his coworker Lonergan at the blast furnace: "the bunch and I took three skirts in Bill's car to Monaca." Certainly revealing, but nothing which pinpoints a place. All of the other evidence in the text is a bit more obscure and requires some knowledge about local history and the history of steelmaking, including that of specific companies.

Many mentions are made about the town in which Walker stays. Near a train station, in a valley, beside a hillside, by the river: All of these are quite generic Pittsburgh area locations. More specifically though, we are in a town that was planned by the company. It is laid out to separate people ethnically and to segregate the management from the workers. This is a powerful tool in our search.

One of the most important leads is found on page 28: "Every furnace protruded a spout, and, when the molten steel inside

AFTERWORD

165

was 'cooked,' tilted backward slightly and poured into a ladle." Although the tilting open hearth was developed in this area, most plants did not use them because they were expensive to build and maintain. Most companies built fixed type open hearth furnaces. (Tilting furnaces were widely used in England and some other European countries because of differences in their supply of raw materials, most specifically the iron ore. This required them to make huge volumes of slag to rid the steel of impurities. The slag could be more easily removed if you could tilt the furnace and leave the molten steel behind until it was completely refined.) There were only three plants in the Pittsburgh area, of the size and complexity Walker describes, that used tilting open hearths.

On his arrival, presumably from the direction of Pittsburgh, he says, "On the right, across tracks, loomed a dark gathering of stacks arising from irregular acres of sheet-iron roofs." Other details that he gives us are about the plant: "Behind them all were five cone shaped towers, against the sky, and a little smoke curling over the top — the blast-furnaces. Behind me the Bessemer furnace threw off a cloud of fire that had changed while I was in the mill from brown to brownish gold." More information is given on pages 96-97, where Adolph the "stove-gang boss" gives Walker a tour: "Each of the other five blast furnaces had a similar lookout and a narrow passageway connecting them with the tops of the stoves.... We could look forth from this eminence and see the whole mill yard, which was nearly a mile in extent...and far to the left were the Bessemers, spouting red gold against a very blue sky. On their right rose the familiar stacks of the open-hearth." One last bit of information will give us all of the data that we need to solve the riddle. On page 129 Walker writes, "The mills had been running for ten years."

So with this we can assemble all of the evidence. The town is one near Pittsburgh and probably Monaca. It's near train tracks and a railroad station which you pass on the way to and from the plant. The plant is on the right, if you assume that you are looking away from Pittsburgh at the train station. The hamlet is situated in a valley that is laid out in a plan that segregates people both ethnically and by position. The plant is one that has five blast furnaces. It makes steel with both Bessemer converters and tilting open hearth furnaces. The layout of the plant is such that if you stand on the bridge of the blast furnace stoves and look to the open hearth furnace shop the Bessemer converters would be on your left. The plant started operating about 1909.

There is only one place that satisfies all of these requirements: the Aliquippa Works of the Jones and Laughlin Steel Company. The plant went into operation in late 1909-10. It had five blast furnaces, tilting open hearth furnaces, and Bessemer converters that were laid out so that the Bessemers were on the left if you were looking at the open hearth shop from the blast furnaces. The town of Aliquippa was started by the Jones and Laughlin Steel Co. (many residents and former residents still refer to areas of town by plan numbers). The Pittsburgh and Lake Erie Railroad train station separated the mill from the community. The plant is on the right of the station if you arrive from Pittsburgh. The next town on the outbound train journey is Monaca. Aliquippa was built in a valley above the mill.

In the process of researching Walker's biographical information, the actual location of the story was confirmed without doubt. The *National Cyclopedia of American Biography* states of Walker, "Following his return to civilian life he [Walker] went to work at the Jones and Laughlin Steel Co., Aliquippa, Pa., as a manual laborer on the open hearth and blast furnaces."

Bouton really is Aliquippa!

Glossary

bath: refers to the mass of molten steel while in the furnace

Bessemer converter (furnace): the pear-shaped vessel in which iron is refined into steel, in a process called a "blow." Invented in 1856 by Henry Bessemer in England and simultaneously by William Kelly of Pittsburgh in Eddyville, KY, the Bessemer process, as it is known, used compressed air blown through a bath of molten iron to reduce the carbon content and refine the iron into steel. The Bessemer process was used at the Aliquippa Works of Jones and Laughlin Steel until 1968.

blast furnace: the tall cone-shaped vessel in which coke and limestone are used to convert iron ore into iron and slag through both temperature and chemical reduction.

bloom: the intermediate product that results from rolling an ingot of steel. The bloom is generally a square shaped bar larger than 5" x 5" in cross section. Smaller than 5" is usually referred to as a billet. Products rolled into rectangular cross section are usually called slabs and circular cross section are rounds. Further rolling is usually required to convert these shapes into useful products. The name "bloom" comes from the fact that the ball of wrought iron, removed from a puddling furnace for further processing, looked like a flower and was called a "blume" (German for "flower").

blooming mill: the place and machinery, i.e. rolls, roll stands, engines, tables, motors, etc., used to convert ingots into smaller shapes called blooms.

blower: the supervisor of operations of a blast furnace.

blue glasses (smoked glasses): furnace workers wore cobalt blue and in older times smoked glasses to reduce the intense glare from the furnace. This enabled them to view the internal workings of the furnace, unimpeded by the bright light.

cast(ing): here, the process of removing the molten iron and slag from a blast furnace.

charging machine: in an open hearth shop, a large electrically driven machine, invented by S.D. Wellman, that is used to fill (charge) the open hearth furnace with the solid constituents (scrap, limestone, iron ore, alloys, etc.) of the steelmaking process. The material is loaded into large rectangular metal (charging) boxes at the stockhouse. The boxes are brought into the furnace shop on long strings of railroad cars and placed in front of the furnace. The charging machine grabs each box, inserts it through the open furnace door, and then rotates it, dumping the contents into the furnace.

checkers (checkerwork): refractory brick, laid in such a way that hot gasses passing over and through heat the brick. Air for combustion in a furnace is then passed through the checkers, removing the stored heat and raising the temperature of the air. Hotter air for combustion makes possible higher temperatures in the furnace. This improvement, known as the regenerative principle when applied to this process of picking up heat from waste gasses, was developed by William Siemens in 1858.

GLOSSARY

coke: the silver-gray material remaining after coal is heated in large chambers (ovens) in the absence of air and essentially all of the volatile material in the coal is driven off. The remains of this process, principally chunks of carbon, is coke. Coke serves a threefold purpose in the blast furnace: first it supports all of the material above it in the furnace (called the burden) while simultaneously letting furnace gasses pass through; second, it acts as a fuel to heat the furnace; third, it acts as a chemical reagent, as a source of carbon, to combine with the oxygen in the iron ore and release (reduce) it as iron.

combustion chamber: in a blast furnace stove, the large open area in which blast furnace and/or coke oven gas is burned, the heat of which is imparted to the checker brick.

company store: a store at which the employees could buy food, clothing, etc. on credit by showing their employee identification. Payment was through payroll deduction. In this case the Jones and Laughlin company store was the Pittsburgh Mercantile Company.

cooler: one of the large cone-shaped water cooled castings in the side of a blast furnace, through which the tuyere was inserted to blow the hot air blast for combustion into the furnace.

dolomite: calcium magnesium carbonate, which is used in the basic lining of an open hearth furnace and used to make repairs to such a furnace.

flue dust: in the case of a blast furnace and stove, the fine particles of iron ore, limestone, coke, carbon, etc. that are carried from the furnace by the gasses flowing through it.

gas house: here, refers to the building where the fuel for the open hearth was made from coal. If large volumes of natural gas or oil were not available or were too expensive, a gas, called producer gas, was made by passing air and steam over red hot coal.

heat: here, refers to one batch of metal in one open hearth furnace, i.e., a heat of steel.

helper: the name given to hourly workers on an open hearth furnace.

> **First helper:** the man in charge of one furnace, usually the most skilled man; was responsible for the furnace.
>
> **Second helper:** the man below the first helper in responsibility; he follows the first helper's directions and had certain duties that included the maintenance of the tap hole and spout, as well as the actual opening of the tap hole.
>
> **Third helper:** junior member of the furnace team; helped with work on more than one furnace; assisted the first and second helpers.

ingot: the solid mass of steel remaining after molten steel was poured or "teemed" into a mold and the mold was removed, after the steel was cooled sufficiently. It is not a final product and is meant for further processing in the mill.

GLOSSARY

jigger: in this case, a Jones and Laughlin Steel term describing a relatively small amount (a number of tons) of iron in a ladle, used to adjust the chemistry of the heat of steel while in the furnace. "Jiggering" (J&L), or the more widely used industry parlance of "pigging up and oreing down," describes a process used by furnace workers to adjust the chemistry of the steel to the desired result. First, a test was made and analyzed. From this information an estimate of the carbon content of the final product would be made. If the carbon would be too low, a calculated amount would be added through the addition of a "jigger" of molten pig iron. Iron has an excess of carbon in comparison to steel, so adding it to the bath would raise the amount of carbon in the steel; hence the term "pigging up." If the amount of carbon in the bath was too high, then raw iron ore would be added to the bath. The oxygen in the iron ore would combine with the carbon in the bath to form carbon monoxide gas, reducing the amount of carbon in the steel; hence the term "oreing down." The process would be repeated as necessary until the desired result was reached.

ladle: a large refractory-lined open-topped cylindrical steel container used to hold molten iron or steel. In some cases, as with iron, it is poured by tilting the molten metal over the top lip of the ladle; or, as in teeming an ingot of steel, it is poured through a small opening in the bottom (the nozzle), which is rigged with a series of levers and rods, capable of stopping the flow of steel when desired.

limestone: calcium carbonate, used to flux (flush) the impurities from both iron and steel and to form a slag.

(ferro)manganese: as used in the text, a material shoveled into a ladle and used to adust the chemistry of the heat of steel while in the ladle. In some cases it was necessary to raise the proportion of manganese to required levels. Sometimes the addition was used to deoxidize the molten metal; to do this was called "killing" the heat.

melter: the man supervising furnace workers in an open hearth shop.

mixer: a large refractory-lined metal vessel used to hold molten iron from a blast furnace before it was charged into a Bessemer converter or open hearth furnace. Frequently as large as 1200 tons in capacity, the mixer was never fully emptied; iron from another cast of a blast furnace was mixed with the remainder, thus providing iron of fairly consistent quality.

mold (mould): the large cast iron form into which molten steel was poured or teemed to make an ingot. Molds could be square, rectangular or round in cross section depending on the final product.

mud: cement-like material made from various constituents (sometimes plain clay), used to seal furnace tap holes or to make repairs to refractory exposed to molten metal outside of the furnace.

mud gun: a steam-driven device used to inject mud into the tap hole of a blast furnace, sealing it off at the end of a cast.

GLOSSARY

open hearth furnace: large rectangular hearth-shaped furnace invented in England in 1858 when William Siemens applied the regenerative principle (the use of checkerwork to recover heat) to a reverbatory furnace. The Martin brothers applied the Siemens invention to melting pig iron and scrap around 1864 in France. "Open hearth" is a term that can be applied to either the acid (the hearth is made of silica brick, ganister, and silica sand which makes an acidic slag) or basic (where the hearth is made of magnesite and dolomite) process. In the basic process the preponderance of phosphorus and sulfur is removed from the metal into the basic slag. In general, the term "Siemens Process" is applied to the acid open hearth and "Siemens-Martin" to the basic open hearth.

pig iron: the iron made in a blast furnace. In the early days of blast furnaces the iron was cast from the runners into rows which had many branches. Early workers said that this arrangement looked like sows with little piglets feeding, hence the name "pig" iron.

pit: in this case the area behind and usually below the open hearth furnaces where they were tapped and the metal was teemed into ingots. The term probably comes from the fact that older furnaces were built at ground level and tapped into ladles set into holes in the ground, or "pits". See photo on page 13.

regenerative principle: see *checkers*.

slag: the molten portion of the bath floating on top of the steel or iron, comprised primarily of limestone added to the process, which included the impurities being flushed from the metal. Some metal is also present.

spout: trough-shaped protrusion from the rear of an open hearth furnace, below the tap hole, through which the molten metal and slag runs into the ladle.

stopper: (in reference to a ladle) the refractory shape seated in the nozzle of a ladle, used by the steel pourer through a series of rods and linkages, to stop and start the flow of molten steel into ingot molds.

stove: in a blast furnace, the tall cylindrical vessels beside the conical blast furnace, which contain checkerwork and a combustion chamber. During one cycle gas (typically blast furnace gas) is burned to heat the checkerwork. During the other cycle, the blast of air for furnace combustion is passed over and through the checkerwork and is heated by the brick so that higher temperatures can be achieved in the blast furnace.

tap: the process of removing molten steel and slag from an open hearth furnace. On fixed open hearth furnaces this was usually done in one of two ways. In both ways some of the seal was dug away from the outside of the tap hole. In one method a large, long rod was pushed by a number of furnace men, through a hole of the door in the front of the furnace, through the molten bath, and through the tap hole from the inside of the furnace (see photo on page 75). In a second method, dynamite was used to blow the tap hole open from the outside. This was called "jet tapping." (Jet tapping had not been invented at the time Walker's book was written). On tilting open hearth furnaces a light plug of mud or a wet plug of burlap was used to close the tap hole. The furnace was tilted backward and the force of the metal would usually push the obstruction out of the way. The hole could not be left open because this would let the slag run into the ladle first, and the slag would then mix with the metal and contaminate the steel.

GLOSSARY

tap drill: the machine used to open the tap hole of a blast furnace so that the furnace may be cast.

tap hole: the hole in the open-hearth furnace or blast furnace, through which the molten metal and sometimes slag was emptied from a furnace, as opposed to the "monkey" or cinder notch on a blast furnace through which only the slag was removed.

teem: the process of pouring molten steel into ingot molds from a ladle.

tuyere: one of the water cooled pipes inserted into the side of a blast furnace through a water cooled hole known as the "cooler." A large pipe, known as the bustle pipe, circles the furnace and distributes the hot blast of air for combustion to each of the tuyeres.

Kenneth J. Kobus has worked in the steel industry for over 30 years and is currently area manager, coal and coke at LTV Steel Co. in Chicago. Both of his grandfathers were immigrant steelworkers; his family has well over 150 years' combined experience at the Pittsburgh Works of Jones and Laughlin/LTV Steel and the Duquesne Works of United States Steel.

He is co-author of *The Pennsy in the Steel City: 150 Years of the Pennsylvania Railroad in Pittsburgh* and of *The Pennsylvania Railroad's Golden Triangle: Main Line Panorama in the Pittsburgh Area* (Pennsylvania Railroad Technical and Historical Society).